Astronomy and General Physics Considered with Reference to Natural Theology

Considered with Reference to Natural Theology

WILLIAM WHEWELL

CAMBRIDGE UNIVERSITY PRESS

Cambridge, New York, Melbourne, Madrid, Cape Town, Singapore,
São Paolo, Delhi, Dubai, Tokyo

Published in the United States of America by Cambridge University Press, New York

www.cambridge.org
Information on this title: www.cambridge.org/9781108000123

This edition first published 1833
This digitally printed version 2009

ISBN 978-1-108-00012-3 Paperback

CAMBRIDGE LIBRARY COLLECTION

Books of enduring scholarly value

Religion

For centuries, scripture and theology were the focus of prodigious amounts of scholarship and publishing, dominated in the English-speaking world by the work of Protestant Christians. Enlightenment philosophy and science, anthropology, ethnology and the colonial experience all brought new perspectives, lively debates and heated controversies to the study of religion and its role in the world, many of which continue to this day. This series explores the editing and interpretation of religious texts, the history of religious ideas and institutions, and not least the encounter between religion and science.

Astronomy and General Physics Considered with Reference to Natural Theology

A leading British intellectual of the Victorian era, William Whewell (1794-1866) was a contemporary and adviser of Herschel, Darwin and Faraday. A geologist, astronomer, theologian, and Master of Trinity College, Cambridge, he was best known for his works on moral philosophy and the history and philosophy of science, and for coining, among others, the term 'scientist'. This book, originally published in 1833, is one of a series of treatises by distinguished authors commissioned with the help of a legacy from the Earl of Bridgewater (d.1829), which were intended to contribute to an understanding of the world as created by God. Though an advocate of religion, Whewell accepts that progress in science leads to an understanding of the laws and processes of the natural world. He argues, however, that ultimately the scientific understanding of creation, astronomy, and the laws of the universe only serves to confirm the idea of a divine designer.

Cambridge University Press has long been a pioneer in the reissuing of out-of-print titles from its own backlist, producing digital reprints of books that are still sought after by scholars and students but could not be reprinted economically using traditional technology. The Cambridge Library Collection extends this activity to a wider range of books which are still of importance to researchers and professionals, either for the source material they contain, or as landmarks in the history of their academic discipline.

Drawing from the world-renowned collections in the Cambridge University Library, and guided by the advice of experts in each subject area, Cambridge University Press is using state-of-the-art scanning machines in its own Printing House to capture the content of each book selected for inclusion. The files are processed to give a consistently clear, crisp image, and the books finished to the high quality standard for which the Press is recognised around the world. The latest print-on-demand technology ensures that the books will remain available indefinitely, and that orders for single or multiple copies can quickly be supplied.

The Cambridge Library Collection will bring back to life books of enduring scholarly value (including out-of-copyright works originally issued by other publishers) across a wide range of disciplines in the humanities and social sciences and in science and technology.

THE BRIDGEWATER TREATISES

ON THE POWER WISDOM AND GOODNESS OF GOD

AS MANIFESTED IN THE CREATION

———

TREATISE III.

ON ASTRONOMY AND GENERAL PHYSICS

BY THE REV. W. WHEWELL.

ET HÆC DE DEO, DE QUO UTIQUE EX PHÆNOMENIS DISSERERE AD
PHILOSOPHIAM NATURALEM PERTINET.

NEWTON, CONCLUSION OF THE PRINCIPIA.

ASTRONOMY AND GENERAL PHYSICS

CONSIDERED WITH REFERENCE TO

NATURAL THEOLOGY.

BY THE

REV. WILLIAM WHEWELL, M.A.

FELLOW AND TUTOR OF TRINITY COLLEGE,

CAMBRIDGE.

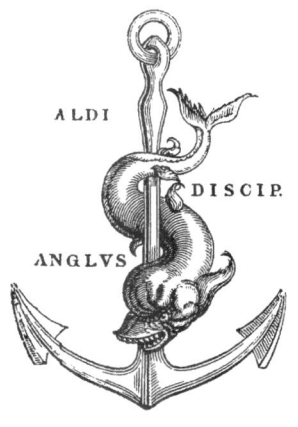

A L D I

D I S C I P.

A N G L V S

LONDON:

WILLIAM PICKERING.

1833.

C. WHITTINGHAM, TOOKS COURT, CHANCERY LANE.

MY LORD,

I owe it to you that I was selected for the task attempted in the following pages, a distinction which I feel to be honourable ; and on this account alone I should have a peculiar pleasure in dedicating the work to your lordship. I do so with additional gratification on another account : the Treatise has been written within the walls of the College of which your lordship was formerly a resident member, and its merits, if it have any, are mainly due to the spirit and habits of the place. The society is always pleased and proud to recollect that a person of the eminent talents and high cha-

racter of your lordship is one of its members ;
and I am persuaded that any effort in the
cause of letters and religion coming from that
quarter, will have for you an interest beyond
what it would otherwise possess.

The subject proposed to me was limited :
my prescribed object is to lead the friends of
religion to look with confidence and pleasure
on the progress of the physical sciences, by
showing how admirably every advance in our
knowledge of the universe harmonizes with
the belief of a most wise and good God. To
do this effectually may be, I trust, a useful
labour. Yet, I feel most deeply, what I would
take this occasion to express, that this, and all
that the speculator concerning Natural Theology can do, is utterly insufficient for the
great ends of Religion ; namely, for the purpose of reforming men's lives, of purifying
and elevating their characters, of preparing
them for a more exalted state of being. It is
the need of something fitted to do this, which
gives to Religion its vast and incomparable

importance ; and this can, I well know, be achieved only by that Revealed Religion of which we are ministers, but on which the plan of the present work did not allow me to dwell.

That Divine Providence may prosper the labours of your lordship, and of all who are joined with you in the task of maintaining and promoting *this* Religion, is, my lord, the earnest wish and prayer of

Your very faithful

and much obliged Servant,

WILLIAM WHEWELL.

Trinity College, Cambridge,
Feb. 25, 1833.

NOTICE.

THE series of Treatises, of which the present is one, is published under the following circumstances:

The RIGHT HONOURABLE and REVEREND FRANCIS HENRY, EARL of BRIDGEWATER, died in the month of February, 1829; and by his last Will and Testament, bearing date the 25th of February, 1825, he directed certain Trustees therein named to invest in the public funds the sum of Eight thousand pounds sterling; this sum, with the accruing dividends thereon, to be held at the disposal of the President, for the time being, of the Royal Society of London, to be paid to the person or persons nominated by him. The Testator further directed, that the person or persons selected by the said President should be appointed to write, print, and publish one thousand copies of a work *On the Power, Wisdom, and Goodness of God, as manifested in the Creation; illustrating such work by all reasonable arguments, as for instance the variety and formation of God's creatures in the animal, vegetable, and mineral kingdoms; the effect of digestion, and thereby of conversion; the construction of the hand of man, and an infinite variety of other arguments; as also by discoveries ancient and modern, in arts, sciences, and the whole extent of literature.* He desired, moreover, that the profits arising from the sale of the works so published should be paid to the authors of the works.

The late President of the Royal Society, Davies Gilbert, Esq. requested the assistance of his Grace the Archbishop of Canterbury and of the Bishop of London, in determining upon the best mode of carrying into effect the intentions of the Testator. Acting with their advice, and with the concurrence of a nobleman immediately connected with the deceased, Mr. Davies Gilbert appointed the following eight gentlemen to write separate Treatises on the different branches of the subject as here stated :

THE REV. THOMAS CHALMERS, D.D.

PROFESSOR OF DIVINITY IN THE UNIVERSITY OF EDINBURGH.

ON THE ADAPTATION OF EXTERNAL NATURE TO THE MORAL AND INTELLECTUAL CONSTITUTION OF MAN.

JOHN KIDD, M.D. F.R.S.

REGIUS PROFESSOR OF MEDICINE IN THE UNIVERSITY OF OXFORD.

ON THE ADAPTATION OF EXTERNAL NATURE TO THE PHYSICAL CONDITION OF MAN.

THE REV. WILLIAM WHEWELL, M.A. F.R.S.

FELLOW OF TRINITY COLLEGE, CAMBRIDGE.

ON ASTRONOMY AND GENERAL PHYSICS.

SIR CHARLES BELL, K.H. F.R.S.

THE HAND: ITS MECHANISM AND VITAL ENDOWMENTS AS EVINCING DESIGN.

PETER MARK ROGET, M.D.

FELLOW OF AND SECRETARY TO THE ROYAL SOCIETY.

ON ANIMAL AND VEGETABLE PHYSIOLOGY.

THE REV. WILLIAM BUCKLAND, D. D. F. R. S.

CANON OF CHRIST CHURCH, AND PROFESSOR OF GEOLOGY IN THE
UNIVERSITY OF OXFORD.

ON GEOLOGY AND MINERALOGY.

THE REV. WILLIAM KIRBY, M. A. F. R. S.

ON THE HISTORY, HABITS, AND INSTINCTS OF ANIMALS.

WILLIAM PROUT, M. D. F. R. S.

ON CHEMISTRY, METEOROLOGY, AND THE FUNCTION OF DIGESTION.

His ROYAL HIGHNESS THE DUKE OF SUSSEX, President of the Royal Society, having desired that no unnecessary delay should take place in the publication of the above mentioned treatises, they will appear at short intervals, as they are ready for publication.

CONTENTS.

[Within the last year or two, several works have been published in this Country on subjects more or less closely approaching to that here treated. It may, therefore, be not superfluous to say that the Author of the following pages believes that he has not borrowed any of his views or illustrations from recent English writers on Natural Theology.]

ASTRONOMY

AND

GENERAL PHYSICS.

INTRODUCTION.

CHAPTER I.

Object of the Present Treatise.

THE examination of the material world brings
before us a number of things and relations of
things which suggest to most minds the belief of
a creating and presiding Intelligence. And this
impression, which arises with the most vague and
superficial consideration of the objects by which
we are surrounded, is, we conceive, confirmed and
expanded by a more exact and profound study
of external nature. Many works have been
written at different times with the view of shew-
ing how our knowledge of the elements and their
operation, of plants and animals and their construction, may serve to nourish and unfold our

idea of a Creator and Governor of the world. But
though this is the case, a new work on the same
subject may still have its use. Our views of the
Creator and Governor of the world, as collected
from or combined with our views of the world
itself, undergo modifications, as we are led by
new discoveries, new generalizations, to regard
nature in a new light. The conceptions concerning the Deity, his mode of effecting his purposes,
the scheme of his government, which are suggested
by one stage of our knowledge of natural objects
and operations, may become manifestly imperfect
or incongruous, if adhered to and applied at a
later period, when our acquaintance with the
immediate causes of natural events has been
greatly extended. On this account it may be
interesting, after such an advance, to shew how
the views of the creation, preservation, and government of the universe, which natural science opens
to us, harmonize with our belief in a Creator,
Governor, and Preserver of the world. To do
this with respect to certain departments of Natural Philosophy is the object of the following pages;
and the author will deem himself fortunate, if
he succeeds in removing any of the difficulties
and obscurities which prevail in men's minds,
from the want of a clear mutual understanding
between the religious and the scientific speculator.
It is needless here to remark the necessarily imperfect and scanty character of Natural Religion;

for most persons will allow that, however imperfect may be the knowledge of a Supreme Intelligence which we gather from the contemplation of the natural world, it is still of most essential use and value. And our purpose on this occasion is, not to shew that Natural Theology is a perfect and satisfactory scheme, but to bring up our Natural Theology to the point of view in which it may be contemplated by the aid of our Natural Philosophy.

Now the peculiar point of view which at present belongs to Natural Philosophy, and especially to the departments of it which have been most successfully cultivated, is, that nature, so far as it is an object of scientific research, is a collection of facts governed by *laws:* our knowledge of nature is our knowledge of laws; of laws of operation and connexion, of laws of succession and co-existence, among the various elements and appearances around us. And it must therefore here be our aim to shew how this view of the universe falls in with our conception of the Divine Author, by whom we hold the universe to be made and governed.

Nature acts by general laws; that is, the occurrences of the world in which we find ourselves, result from causes which operate according to fixed and constant rules. The succession of days, and seasons, and years, is produced by the motions of the earth ; and these again are governed by the attraction of the sun, a force which acts with undeviating steadiness and regularity. The changes of winds and skies, seemingly so capricious and casual, are produced by the operation of the sun s heat upon air and moisture, land and sea ; and though in this case we cannot trace the particular events to their general causes, as we can trace the motions of the sun and moon, no philosophical mind will doubt the generality and fixity of the rules by which these causes act. The variety of the effects takes place, because the circumstances in different cases vary ; and not because the action of material causes leaves anything to chance in the result. And again, though the vital movements which go on in the frame of vegetables and animals depend on agencies still less known, and probably still more complex, than those which rule the weather, each of the powers on which such movements depend has its peculiar laws of action, and these are as universal and as invariable as the law by which a stone falls to the earth when not supported.

The world then is governed by general laws ; and in order to collect from the world itself a judgment concerning the nature and character of its government, we must consider the import and tendency of such laws, so far as they come under our knowledge. If there be, in the administration of the universe, intelligence and benevolence, superintendence and foresight, grounds for love and hope, such qualities may be expected to appear in the constitution and combination of those fundamental regulations by which the course of nature is brought about, and made to be what it is.

If a man were, by some extraordinary event, to find himself in a remote and unknown country, so entirely strange to him that he did not know whether there existed in it any law or government at all; he might in no long time ascertain whether the inhabitants were controlled by any superintending authority; and with a little attention he might determine also whether such authority were exercised with a prudent care for the happiness and well-being of its subjects, or without any regard and fitness to such ends; whether the country were governed by laws at all, and whether the laws were good. And according to the laws which he thus found prevailing, he would judge of the sagacity, and the purposes of the legislative power.

By observing the laws of the material universe and their operation, we may hope, in a somewhat similar manner, to be able to direct our judgment concerning the government of the universe: concerning the mode in which the elements are regulated and controlled, their effects combined and balanced. And the general tendency of the results thus produced may discover to us something of the character of the power which has legislated for the material world.

We are not to push too far the analogy thus suggested. There is undoubtedly a wide difference between the circumstances of man legislating for man, and God legislating for matter. Still we shall, it will appear, find abundant reason to admire the wisdom and the goodness which have established *the Laws of Nature*, however rigorously we may scrutinize the import of this expression.

CHAPTER II.

On Laws of Nature.

WHEN we speak of material nature as being governed by *laws*, it is sufficiently evident that we use the term in a manner somewhat metaphorical. The laws to which man's attention is primarily directed are *moral* laws; rules laid down for his actions; rules for the conscious actions of a person; rules which, as a matter of possibility, he may obey or may transgress; the latter event being combined, not with an impossibility, but with a penalty. But the *Laws of Nature* are something different from this; they are rules for that which *things* are to do and suffer; and this by no consciousness or will of theirs. They are rules describing the mode in which things *do* act; they are invariably obeyed ; their transgression is not punished, it is excluded. The language of a moral law is, man *shall* not kill ; the language of a Law of Nature is, a stone *will* fall to the earth.

These two kinds of laws direct the actions of persons and of things, by the sort of control of which persons and things are respectively susceptible ; so that the metaphor is very simple ; but it is proper for us to recollect that it is a metaphor, in order that we may clearly apprehend what is implied in speaking of the Laws of Nature.

In this phrase are included all properties of the portions of the material world ; all modes of action and rules of causation, according to which they operate on each other. The whole course of the visible universe therefore is but the collective result of such laws ; its movements are only the aggregate of *their* working. All natural occurrences, in the skies and on the earth, in the organic and in the inorganic world, are determined by the relations of the elements and the actions of the forces of which the rules are thus prescribed.

The relations and rules by which these occurrences are thus determined necessarily depend on measures of time and space, motion and force ; on quantities which are subject to numerical measurement, and capable of being connected by mathematical properties. And thus all things are ordered by number and weight and measure. " God," as was said by the ancients, " works by geometry :" the legislation of the material universe is necessarily delivered in the language of mathematics ; the stars in their courses are regulated by the properties of conic sections, and the winds depend on arithmetical and geometrical progressions of elasticity and pressure.

The constitution of the universe, so far as it can be clearly apprehended by our intellect, thus assumes a shape involving an assemblage of mathematical propositions : certain algebraical formulæ, and the knowledge when and how to apply them, constitute the last step of the physical science to which we can attain. The labour and the endowments of ages have been employed in bringing such science into the condition in which it now exists ; and an exact and extensive discipline in mathematics, followed by a practical and profound study of the researches of natural philosophers, can alone put any one in possession of all the knowledge concerning the course of the material world, which is at present open to man. The general impression, however, which arises from the view thus obtained of the universe, the results which we collect from the most careful scrutiny of its administration, may, we trust, be rendered intelligible without this technical and laborious study, and to do this is our present object.

It will be our business to shew that the laws which really prevail in nature are, by their *form*, that is, by the nature of the connexion which they establish among the quantities and properties which they regulate, remarkably adapted to the office which is assigned them ; and thus offer evidence of selection, design, and goodness, in the power by which they were established. But these characters of the legislation of the universe may also be seen, in many instances, in a manner somewhat different from the selection of the law The *nature of the connexion* remaining the same, the quantities which it regulates may also in their *magnitude* bear marks of selection and purpose. For the law may be the same while the quantities to which it applies are different. The law of the gravity which acts to the earth and to Jupiter, is the same ; but the intensity of the force at the surfaces of the two planets is different. The law which regulates the density of the air at any point, with reference to the height from the earth's surface, would be the same, if the atmosphere were ten times as large, or only one tenth as large as it is ; if the barometer at the earth's surface stood at three inches only, or if it shewed a pressure of thirty feet of mercury

Now this being understood, the adaptation of a law to its purpose, or to other laws, may appear in two ways :—either in the form of the law, or in the amount of the magnitudes which it regulates, which are sometimes called *arbitrary magnitudes*.

If the attraction of the sun upon the planets did not vary inversely as the square of the distance, the *form* of the law of gravitation would be changed; if this attraction were, at the earth's orbit, of a different *value* from its present one, the arbitrary magnitude would be changed; and it will appear, in a subsequent part of this work, that either change would, so far as we can trace its consequences, be detrimental. The form of the law determines in what manner the facts shall take place; the arbitrary magnitude determines how fast, how far, how soon; the one gives a model, the other a measure of the phenomenon; the one draws the plan, the other gives the scale on which it is to be executed; the one gives the rule, the other the rate. If either were wrongly taken, the result would be wrong too.

CHAPTER III.

Mutual Adaptation in the Laws of Nature.

To ascertain such laws of nature as we have been describing, is the peculiar business of science. It is only with regard to a very small portion of the appearances of the universe, that science, in any strict application of the term, exists. In very few departments of research have men been able to trace a multitude of known facts to causes which appear to be the ultimate material causes, or to discern the laws which seem to be the most general laws. Yet, in one or two instances, they have done this, or something approaching to this; and most especially in the instance of that part of nature, which it is the object of this treatise more peculiarly to consider.

The apparent motions of the sun, moon, and stars have been more completely reduced to their causes and laws than any other class of phenomena. Astronomy, the science which treats of these, is already a wonderful example of the degree of such knowledge which man may attain. The forms of its most important laws may be conceived to be certainly known; and hundreds of observers in all parts of the world are daily employed in determining, with additional accuracy,

the arbitrary magnitudes which these laws involve.

The enquiries in which the mutual effects of
heat, moisture, air, and the like elements are
treated of, including, among other subjects, all
that we know of the causes of the weather (meteorology) is a far more imperfect science than
astronomy. Yet, with regard to these agents, a
great number of laws of nature have been discovered, though, undoubtedly, a far greater number remain still unknown.

So far, therefore, as our knowledge goes, astronomy and meteorology are parts of natural
philosophy in which we may study the order of
nature with such views as we have suggested ;
in which we may hope to make out the adaptations and aims which exist in the laws of nature ;
and thus to obtain some light on the tendency of
this part of the legislation of the universe, and on
the character and disposition of the Legislator.

The number and variety of the laws which we
find established in the universe is so great, that
it would be idle to endeavour to enumerate them.
In their operation they are combined and intermixed in incalculable and endless complexity,
influencing and modifying each other s effects in
every direction. If we attempt to comprehend
at once the whole of this complex system, we find
ourselves utterly baffled and overwhelmed by its
extent and multiplicity. Yet, in so far as we

consider the bearing of one part upon another, we receive an impression of adaptation, of mu tual fitness, of conspiring means, of preparation and completion, of purpose and provision. This impression is suggested by the contemplation of every part of nature ; but the grounds of it, from the very circumstances of the case, cannot be conveyed in a few words. It can only be fully educed by leading the reader through several views and details, and must grow out of the combined influence of these on a sober and reflecting frame of mind. However strong and solemn be the conviction which may be derived from a contemplation of nature, concerning the existence, the power, the wisdom, the goodness of our Divine Governor, we cannot expect that this conviction, as resulting from the extremely complex spectacle of the material world, should be capable of being irresistibly conveyed by a few steps of reasoning, like the conclusion of a geometrical proposition, or the result of an arithmetical calculation.

We shall, therefore, endeavour to point out cases and circumstances in which the different parts of the universe exhibit this mutual adaptation, and thus to bring before the mind of the reader the evidence of wisdom and providence, which the external world affords. When we have illustrated the correspondencies which exist in every province of nature, between the qualities of brute matter and the constitution of living things, between the tendency to derangement and the conservative influences by which such a tendency is counteracted, between the office of the minutest speck and of the most general laws ; it will, we trust, be difficult or impossible to exclude from our conception of this wonderful system, the idea of a harmonizing, a preserving, a contriving, an intending Mind ; of a Wisdom, Power, and Goodness far exceeding the limits of our thoughts.

CHAPTER IV.

Division of the Subject.

In making a survey of the universe, for the purpose of pointing out such correspondencies and adaptations as we have mentioned, we shall suppose the general leading facts of the course of nature to be known, and the explanations of their causes now generally established among astronomers and natural philosophers to be conceded. We shall assume therefore that the earth is a solid globe of ascertained magnitude, which travels round the sun, in an orbit nearly circular, in a period of about three hundred and sixty five days and a quarter, and in the mean time revolves, in an inclined position, upon its own axis in about twenty-four hours, thus producing the succession of appearances and effects which constitute seasons and climates, day and night;—that this globe has its surface furrowed and ridged with various inequalities, the waters of the ocean occupying the depressed parts:—that it is surrounded by an atmosphere, or spherical covering of air; and that various other physical agents, moisture, electricity, magnetism, light, operate at the surface of the earth, according to their peculiar laws. This surface is, as we know, clothed with a covering of plants, and inhabited by the various tribes of animals, with all their variety of sensations, wants, and enjoyments. The relations and connexions of the larger portions of the world, the sun, the planets, and the stars, the *cosmical* arrangements of the system, as they are sometimes called, determine the course of events among these bodies; and the more remarkable features of these arrangements are therefore some of the subjects for our consideration. These cosmical arrangements, in their consequences, affect also the physical agencies which are at work at the surface of the earth, and hence come in contact with *terrestrial* occurrences. They thus influence the functions of plants and animals. The circumstances in the cosmical system of the universe, and in the organic system of the earth, which have thus a bearing on each other, form another of the subjects of which we shall treat. The former class of considerations attends principally to the stability and other apparent perfections of the solar system;

the latter to the well being of the system of organic life by which the earth is occupied. The two portions of the subject may be treated as Cosmical Arrangements and Terrestrial Adaptations.

We shall begin with the latter class of adaptations, because in treating of these the facts are more familiar and tangible, and the reasonings less abstract and technical, than in the other division of the subject. Moreover, in this case men have no difficulty in recognizing as desirable the end which is answered by such adaptations, and they therefore the more readily consider it *as an end*. The nourishment, the enjoyment, the diffusion of living things, are willingly acknowledged to be a suitable object for contrivance ; the simplicity, the permanence, of an inert mechanical combination might not so readily be allowed to be a manifestly worthy aim of a Creating Wisdom. The former branch of our argument may therefore be best suited to introduce to us the Deity as the institutor of Laws of Nature, though the latter may afterwards give us a wider view and a clearer insight into one province of his legislation.

BOOK I.

TERRESTRIAL ADAPTATIONS.

WE proceed in this Book to point out relations which subsist between the laws of the inorganic world, that is, the general facts of astronomy and meteorology ; and the laws which prevail in the organic world, the properties of plants and animals.

With regard to the first kind of laws, they are in the highest degree various and unlike each other. The intensity and activity of natural influences follow in different cases the most different rules. In some instances they are *periodical*, increasing and diminishing alternately, in a perpetual succession of equal intervals of time. This is the case with the heat at the earth's surface, which has a period of a year ; with the light, which has a period of a day   Other qualities are *constant*, thus the force of gravity at the same place is always the same. In some cases, a very simple cause produces very complicated effects ; thus the globular form of the earth, and the inclination of its axis during its annual motion, give rise to all the variety of climates. In other cases

C

a very complex and variable system of causes
produces effects comparatively steady and uniform ; thus solar and terrestrial heat, air, moisture, and probably many other apparently con-
flicting agents, join to produce our weather, which
never deviates very far from a certain average
standard.

Now a general fact, which we shall endeavour to
exemplify in the following chapters, is this:—That
those properties of plants and animals which have
reference to agencies of a periodical character,
have also by their nature a periodical mode of
working ; while those properties which refer to
agencies of constant intensity, are adjusted to this
constant intensity : and again, there are peculiarities in the nature of organized beings which
have reference to a variety in the conditions of the
external world, as, for instance, the difference of
the organized population of different regions : and
there are other peculiarities which have a reference to the constancy of the average of such
conditions, and the limited range of the deviations
from that average ; as for example, that constitution by which each plant and animal is fitted to
exist and prosper in its usual place in the world.

And not only is there this general agreement
between the nature of the laws which govern the
organic and inorganic world, but also there is a
coincidence between the *arbitrary magnitudes*
which such laws involve on the one hand and on

the other. Plants and animals have, in their construction, certain periodical functions, which have
a reference to alternations of heat and cold ; the
length of the period which belongs to these functions by their construction, appears to be that of
the period which belongs to the actual alternations
of heat and cold, namely, a year.   Plants and
animals have again in their construction certain
other periodical functions, which have a reference
to alternations of light and darkness ; the length of
the period of such functions appears to coincide
with the natural day   In like manner the other
arbitrary magnitudes which enter into the laws
of gravity, of the effects of air and moisture, and
of other causes of permanence, and of change, by
which the influences of the elements operate, are
the same arbitrary magnitudes to which the members of the organic world are adapted by the
various peculiarities of their construction.

The illustration of this view will be pursued in
the succeeding chapters ; and when the coincidence here spoken of is distinctly brought before
the reader, it will, we trust, be found to convey
the conviction of a wise and benevolent design,
which has been exercised in producing such an
agreement between the internal constitution and
the external circumstances of organized beings.
We shall adduce cases where there is an apparent
relation between the course of operation of the
elements and the course of vital functions ; be-

tween some fixed measure of time or space, traced in the lifeless and in the living world; where creatures are constructed on a certain plan, or a certain scale, and this plan or this scale is exactly the single one which is suited to their place on the earth; where it was necessary for the Creator (if we may use such a mode of speaking) *to take account* of the weight of the earth, or the density of the air, or the measure of the ocean, and where these quantities are rightly taken account of in the arrangements of creation. In such cases we conceive that we trace a Creator, who, in producing one part of his work, was not forgetful or careless of another part; who did not cast his living creatures into the world to prosper or perish as they might find it suited to them or not; but fitted together, with the nicest skill, the world and the constitution which he gave to its inhabitants; so fashioning it and them, that light and darkness, sun and air, moist and dry, should become their ministers and benefactors, the unwearied and unfailing causes of their well being.

We have spoken of the mutual adaptation of the organic and the inorganic world. If we were to conceive the contrivance of the world as taking place in an order of time in the contriving mind, we might also have to conceive this adaptation as taking place in one of two ways: we might either suppose the laws of inert nature to be accommodated to the foreseen wants of living things, or the organization of life to be accommodated to the previously established laws of nature. But we are not forced upon any such mode of conception, or upon any decision between such suppositions : since, for the purpose of our argument, the consequence of either view is the same. There is an adaptation somewhere or other, on either supposition. There is account taken of one part of the system in framing the other : and the mind which took such account can be no other than that of the Intelligent Author of the universe. When indeed we come to see the vast number, the variety, the extent, the interweaving, the reconciling of such adaptations, we shall readily allow, that all things are so moulded upon and locked into each other, connected by such subtilty and profundity of design, that we may well abandon the idle attempt to trace the *order* of thought in the mind of the Supreme Ordainer.

CHAPTER I.

The Length of the Year.

A YEAR is the most important and obvious of the periods which occur in the organic, and especially in the vegetable world. In this interval of time the cycle of most of the external influences which

operate upon plants is completed.   There is also
in plants a cycle of internal functions, corresponding to this succession of external causes.
The length of either of these periods might have
been different from what it is, according to any
grounds of necessity which we can perceive.
But a certain length is selected in both instances,
and in both instances the same.   The length of
the year is so determined as to be adapted to the
constitution of most vegetables ; or the construction of vegetables is so adjusted as to be suited to
the length which the year really has, and unsuited
to a duration longer or shorter by any considerable portion.   The vegetable clock-work is so set
as to go for a year.

The length of the year or interval of recurrence
of the seasons is determined by the time which
the earth employs in performing its revolution
round the sun : and we can very easily conceive
the solar system so adjusted that the year should
be longer or shorter than it actually is.   We can
imagine the earth to revolve round the sun at a
distance greater or less than that which it at
present has, all the forces of the system remaining
unaltered.   If the earth were removed towards
the centre by about one-eighth of its distance, the
year would be diminished by about a month ; and
in the same manner it would be increased by a
month on increasing the distance by one-eighth.
We can suppose the earth at a distance of 84 or

108 millions of miles, just as easily as at its present distance of 96 millions : we can suppose the earth with its present stock of animals and vegetables placed where Mars or where Venus is, and revolving in an orbit like one of theirs : on the former supposition our year would become twenty-three, on the latter seven of our present months. Or we can conceive the present distances of the parts of the system to continue what they are, and the size, or the density of the central mass, the sun, to be increased or diminished in any proportion ; and in this way the time of the earth s revolution might have been increased or diminished in any degree ; a greater velocity, and consequently a diminished period, being requisite, in order to balance an augmented central attraction. In any of these ways the length of the earth's natural year might have been different from what it now is : in the last way without any necessary alteration, so far as we can see, of temperature.

Now, if any change of this kind were to take place, the working of the botanical world would be thrown into utter disorder, the functions of plants would be entirely deranged, and the whole vegetable kingdom involved in instant decay and rapid extinction.

That this would be the case, may be collected from innumerable indications. Most of our fruit trees, for example, require the year to be of its

present length. If the summer and the autumn
were much shorter, the fruit could not ripen ; if
these seasons were much longer, the tree would
put forth a fresh suit of blossoms, to be cut down
by the winter. Or if the year were twice its
present length, a second crop of fruit would
probably not be matured, for want, among other
things, of an intermediate season of rest and
consolidation, such as the winter is. Our forest
trees in like manner appear to need all the seasons
of our present year for their perfection ; the
spring, summer, and autumn, for the developement of their leaves and consequent formation of
their *proper juice*, and of wood from this ; and the
winter for the hardening and solidifying the
substance thus formed.

Most plants, indeed, have some peculiar function adapted to each period of the year, that is of
the now existing year. The sap ascends with
extraordinary copiousness at two seasons, in the
spring and in the autumn, especially the former.
The opening of the leaves and the opening of the
flowers of the same plants are so constant to their
times, (their *appointed* times, as we are naturally
led to call them,) that such occurrences might be
taken as indications of the times of the year. It
has been proposed in this way to select a series
of botanical facts which should form a calendar ;
and this has been termed a *calendar of Flora*.
Thus, if we consider the time of putting forth

leaves,* the honeysuckle protrudes them in the month of January ; the gooseberry, currant, and elder in the end of February, or beginning of March ; the willow, elm, and lime-tree in April ; the oak and ash, which are always the latest among trees, in the beginning or towards the middle of May. In the same manner the flowering has its regular time : the mezereon and snowdrop push forth their flowers in February ; the primrose in the month of March ; the cowslip in April ; the great mass of plants in May and June; many in July, August, and September ; some not till the month of October, as the meadow saffron; and some not till the approach and arrival of winter, as the laurustinus and arbutus.

The fact which we have here to notice, is the recurrence of these stages in the developement of plants, at intervals precisely or very nearly of twelve months. Undoubtedly, this result is in part occasioned by the action of external stimulants upon the plant, especially heat, and by the recurrence of the intensity of such agents. Accordingly, there are slight differences in the times of such occurrences, according to the backwardness or forwardness of the season, and according as the climate is genial or otherwise. Gardeners use artifices which will, to a certain extent, accelerate or retard the time of developement of a

* Loudon, Encyclopædia of Gardening, 848.

plant. But there are various circumstances which shew that this recurrence of the same events and equal intervals is not entirely owing to external causes, and that it depends also upon something in the internal structure of vegetables. Alpine plants do not wait for the stimulus of the sun s heat, but exert such a struggle to blossom, that their flowers are seen among the yet unmelted snow. And this is still more remarkable in the naturalization of plants from one hemisphere to the other When we transplant our fruit trees to the temperate regions south of the equator, they continue for some years to flourish at the period which corresponds to our spring. The reverse of this obtains, with certain trees of the southern hemisphere. Plants from the Cape of Good Hope, and from Australia, countries whose summer is simultaneous with our winter, exhibit their flowers in the coldest part of the year, as the heaths.

This view of the subject agrees with that maintained by the best Botanical writers. Thus Decandolle observes that after making allowance for all meteorological causes, which determine the epoch of flowering, we must reckon as another cause the peculiar nature of each species. The flowering once determined, appears to be subject to a law of *periodicity* and habit.*

* Dec. Phys. vol. ii. 478.

It appears then that the functions of plants have by their nature a periodical character ; and the length of the period thus belonging to vegetables is a result of their organization. Warmth and light, soil and moisture, may in some degree modify, and hasten or retard the stages of this period ; but when the constraint is removed the natural period is again resumed. Such stimulants as we have mentioned are not the *causes* of this periodicity They do not produce the varied functions of the plant, and could not occasion their performance at regular intervals, except the plant possessed a suitable construction. They could not alter the length of the cycle of vegetable functions, except within certain very narrow limits. The processes of the rising of the sap, of the formation of proper juices, of the unfolding of leaves, the opening of flowers, the fecundation of the fruit, the ripening of the seed, its proper deposition in order for the reproduction of a new plant ;—all these operations require a certain portion of time, and could not be compressed into a space less than a year, or at least could not be abbreviated in any very great degree. And on the other hand, if the winter were greatly longer than it now is, many seeds would not germinate at the return of spring. Seeds which have been kept too long require stimulants to make them fertile.

If therefore the duration of the seasons were

much to change, the processes of vegetable life would be interrupted, deranged, distempered. What, for instance, would become of our calendar of Flora, if the year were lengthened or shortened by six months? Some of the dates would never arrive in the one case, and the vegetable processes which mark them would be superseded; some seasons would be without dates in the other case, and these periods would be employed in a way harmful to the plants, and no doubt speedily destructive. We should have not only *a year of confusion*, but, if it were repeated and continued, a year of death.

But in the existing state of things, the duration of the earth's revolution round the sun, and the duration of the revolution of the vegetable functions of most plants are equal. These two periods are *adjusted* to each other. The stimulants which the elements apply come at such intervals and continue for such times, that the plant is supported in health and vigour, and enabled to reproduce its kind. Just such a portion of time is measured out for the vegetable powers to execute their task, as enables them to do so in the best manner.

Now such an adjustment must surely be accepted as a proof of design, exercised in the formation of the world. Why should the solar year be so long and no longer? or, this being of such a length, why should the vegetable cycle be ex-

actly of the same length ? Can this be chance ?
And this occurs, it is to be observed, not in one,
or in a few species of plants, but in thousands.
Take a small portion only of known species, as
the most obviously endowed with this adjustment,
and say ten thousand. How should all these organized bodies be constructed for the same period
of a year. How should all these machines be
wound up so as to go for the same time ? Even
allowing that they could bear a year of a month
longer or shorter, how do they all come within
such limits ? No chance could produce such a
result. And if not by chance, how otherwise
could such a coincidence occur, than by an intentional adjustment of these two things to one
another ? by a selection of such an organization
in plants, as would fit them to the earth on which
they were to grow ; by an adaptation of construction to conditions ; of the scale of the construction
to the scale of the conditions.

It cannot be accepted as an explanation of this
fact in the economy of plants, that it is necessary to
their existence ; that no plants could possibly have
subsisted, and come down to us, except those
which were thus suited to their place on the earth.
This is true ; but this does not at all remove the
necessity of recurring to design as the origin
of the construction by which the existence and
continuance of plants is made possible. A watch
could not go, except there were the most exact

adjustment in the forms and positions of its
wheels ; yet no one would accept it as an explanation of the origin of such forms and positions,
that the watch would not go if these were other
than they are.   If the objector were to suppose
that plants were originally fitted to years of various lengths, and that such only have survived to
the present time, as had a cycle of a length equal
to our present year, or one which could be accommodated to it ; we should reply, that the assumption is too gratuitous and extravagant to require
much consideration ; but that, moreover, it does
not remove the difficulty.   How came the functions of plants to be *periodical* at all ?   Here is,
in the first instance, an agreement in the form of
the laws that prevail in the organic and in the
inorganic world, which appears to us a clear
evidence of design in their Author.   And the
same kind of reply might be made to any similar
objection to our argument.   Any supposition
that the universe has gradually approximated to
that state of harmony among the operations of its
different parts, of which we have one instance in
the coincidence now under consideration, would
make it necessary for the objector to assume a
previous state of things preparatory to this perfect correspondence.   And in this preparatory
condition we should still be able to trace the ru-
diments of that harmony, for which it was proposed to account: so that even the most unbounded

license of hypothesis would not enable the oppo-
nent to obliterate the traces of an intentional
adaptation of one part of nature to another.

Nor would it at all affect the argument, if these
periodical occurrences could be traced to some
proximate cause: if for instance it could be
shewn, that the budding or flowering of plants
is brought about at particular intervals, by the
nutriment accumulated in their vessels during
the preceding months. For the question would
still remain, how their functions were so adjusted,
that the accumulation of the nutriment necessary
for budding and flowering, together with the
operation itself, comes to occupy exactly a year,
instead of a month only, or ten years. There
must be in their structure some reference to time:
how did such a reference occur? how was it
determined to the particular time of the earth's
revolution round the sun? This could be no
otherwise, as we conceive, than by design and
appointment.

We are left therefore with this manifest adjustment before us, of two parts of the universe, at
first sight so remote; the dimensions of the solar
system and the powers of vegetable life. These
two things are so related, that one has been made
to fit the other. The relation is as clear as that
of a watch to a sundial. If a person were to compare the watch with the dial, hour after hour,
and day after day, it would be impossible for him

not to believe that the watch had been *contrived* to accommodate itself to the solar day. We have at least ten thousand kinds of vegetable watches of the most various forms, which are all accommodated to the solar year ; and the evidence of contrivance seems to be no more capable of being eluded in this case than in the other.

The same kind of argument might be applied to the animal creation. The pairing, nesting, hatching, fledging, and flight of birds, for instance, occupy each its peculiar time of the year ; and, together with a proper period of rest, fill up the twelve months. The transformations of most insects have a similar reference to the seasons, their progress and duration. " In every species" (except man), says a writer* on animals, " there is a particular period of the year in which the reproductive system exercises its energies. And the season of love and the period of gestation are so arranged that the young ones are produced at the time wherein the conditions of temperature are most suited to the commencement of life." It is not our business here to consider the details of such provisions, beautiful and striking as they are. But the prevalence of the great law of periodicity in the vital functions of organized beings will be allowed to have a claim to be considered in its reference to astronomy, when it is

* Fleming, Zool. i. 400.

seen that their periodical constitution derives its use from the periodical nature of the motions of the planets round the sun ; and that the duration of such cycles in the existence of plants and animals has a reference to the arbitrary elements of the solar system : a reference which, we maintain, is inexplicable and unintelligible, except by admitting into our conceptions an Intelligent Author, alike of the organic and inorganic universe.

Chapter II.

The Length of the Day.

WE shall now consider another astronomical element, the time of the revolution of the earth on its axis ; and we shall find here also that the structure of organized bodies are suited to this element ;---that the cosmical and physiological arrangements are adapted to each other.

We can very easily conceive the earth to revolve on her axis faster or slower than she does, and thus the days to be longer or shorter than they are, without supposing any other change to take place. There is no apparent reason why this globe should turn on its axis just three hundred and sixty-six times while it describes its orbit round the sun. The revolutions of the other planets, so far as we know them, do not

appear to follow any rule by which they are connected with the distance from the sun. Mercury,
Venus, and Mars have days nearly the length of
ours. Jupiter and Saturn revolve in about ten
hours each. For any thing we can discover, the
earth might have revolved in this or any other
smaller period; or we might have had, without
mechanical inconvenience, much longer days than
we have.

But the terrestrial day, and consequently the
length of the cycle of light and darkness, being
what it is, we find various parts of the constitution both of animals and vegetables, which have
a periodical character in their functions, corresponding to the diurnal succession of external
conditions; and we find that the length of the
period, as it exists in their constitution, coincides
with the length of the natural day.

The alternation of processes which takes place
in plants by day and by night is less obvious,
and less obviously essential to their well-being,
than the annual series of changes. But there are
abundance of facts which serve to shew that such
an alternation is part of the vegetable economy.

In the same manner in which Linnæus proposed a Calendar of Flora, he also proposed a
Dial of Flora, or Flower-Clock; and this was to
consist, as will readily be supposed, of plants,
which mark certain hours of the day, by opening
and shutting their flowers. Thus the day-lily

(hemerocallis fulva) opens at five in the morning ; the *leontodon taraxacum*, or common dandelion, at five or six ; the *hieracium latifolium* (hawk-weed), at seven ; the *hieracium pilosella*, at eight; the *calendula arvensis*, or marigold, at nine ; the *mesembryanthemum neapolitanum*, at ten or eleven : and the closing of these and other flowers in the latter part of the day offers a similar system of hour marks.

Some of these plants are thus expanded in consequence of the stimulating action of the light and heat of the day, as appears by their changing their time, when these influences are changed ; but others appear to be constant to the same hours, and independent of the impulse of such external circumstances.   Other flowers by their opening and shutting prognosticate the weather. Plants of the latter kind are called by Linnæus, *meteoric* flowers, as being regulated by atmospheric causes : those which change their hour of opening and shutting with the length of the day, he terms *tropical;* and the hours which they measure are, he observes, like Turkish hours, of varying length at different seasons.   But there are other plants which he terms *equinoctial;* their vegetable days, like the days of the equator, being always of equal length ; and these open, and generally close, at a fixed and positive hour of the day.   Such plants clearly prove that the periodical character, and the period of the motions above described, do not depend altogether on external circumstances.

Some curious experiments on this subject were made by Decandolle. He kept certain plants in two cellars, one warmed by a stove and dark, the other lighted by lamps. On some of the plants the artificial light appeared to have no influence, (*convolvulus arvensis, convolvulus cneorum, silene fruticosa*) and they still followed the clock hours in their opening and closing. The night-blowing plants appeared somewhat disturbed, both by perpetual light and perpetual darkness. In either condition they accelerated their *going* so much, that in three days they had gained half a day, and thus exchanged night for day as their time of opening. Other flowers *went slower* in the artificial light (*convolvulus purpureus*). In like manner those plants which fold and unfold their leaves were variously effected by this mode of treatment. The *oxalis stricta* and *oxalis incarnata* kept their habits, without regarding either artificial light or heat. The *mimosa leucocephala* folded and unfolded at the usual times, whether in light or in darkness, but the folding up was not so complete as in the open air. The *mimosa pudica* (sensitive plant), kept in darkness during the day time, and illuminated during the night, had in three days accommodated herself to the artificial state, opening in the evening, and closing in the morning ; restored to the open air, she recovered her usual habits.

Tropical plants in general, as is remarked by our gardeners, suffer from the length of our summer daylight ; and it has been found necessary to shade them during a certain part of the day.

It is clear from these facts, that there is a diurnal period belonging to the constitution of vegetables ; though the succession of functions depends in part on external stimulants, as light and heat, their periodical character is a result of the structure of the plant ; and this structure is such, that the length of the period, under the common influences to which plants are exposed, coincides with the astronomical day. The power of accommodation which vegetables possess in this respect, is far from being such as either to leave the existence of this periodical constitution doubtful, or to entitle us to suppose that the day might be considerably lengthened or shortened without injury to the vegetable kingdom.

Here then we have an adaptation between the structure of plants, and the periodical order of light and darkness which arises from the earth's rotation ; and the arbitrary quantity, the length of the cycle of the physiological and of the astronomical fact, is the same. Can this have occurred any otherwise than by an intentional adjustment?

Any supposition that the astronomical cycle has occasioned the physiological one, that the

structure of plants has been brought to be what it
is by the action of external causes, or that such
plants as could not accommodate themselves to
the existing day have perished, would be not only
an arbitrary and baseless assumption, but moreover useless for the purposes of explanation which
it professes, as we have noticed of a similar supposition with respect to the annual cycle.   How
came plants to have periodicity at all in those
functions which have a relation to light and darkness ?   This part of their constitution was suited
to organised things which were to flourish on the
earth, and it is accordingly bestowed on them ; it
was necessary for this end that the period should
be of a certain length ; it is of that length and no
other.   Surely this looks like intentional provision.

Animals also have a period in their functions and
habits ; as in the habits of waking, sleeping, eating,
&c. and their well-being appears to depend on the
coincidence of this period with the length of the
natural day.   We see that in the day, as it now
is, all animals find seasons for taking food and
repose, which agree perfectly with their health
and comfort.   Some animals feed during the day,
as nearly all the ruminating animals and land
birds ; others feed only in the twilight, as bats
and owls, and are called *crepuscular ;* while many
beasts of prey, aquatic birds, and others, take
their food during the night.   Those animals which
are nocturnal feeders are diurnal sleepers, while

those which are crepuscular, sleep partly in the
night and partly in the day ; but in all, the complete period of these functions is twenty-four
hours. Man, in like manner, in all nations and
ages, takes his principal rest once in twenty-four
hours ; and the regularity of this practice seems
most suitable to his health, though the duration
of the time allotted to repose is extremely different in different cases. So far as we can judge, this
period is of a length beneficial to the human frame,
independently of the effect of external agents. In
the voyages recently made into high northern latitudes, where the sun did not rise for three months,
the crews of the ships were made to adhere, with
the utmost punctuality, to the habit of retiring to
rest at nine, and rising a quarter before six ; and
they enjoyed, under circumstances apparently the
most trying, a state of salubrity quite remarkable.
This shews, that according to the common constitution of such men, the cycle of twenty-four
hours is very commodious, though not imposed
on them by external circumstances.

The hours of food and repose are capable of
such wide modifications in animals, and above all
in man, by the influence of external stimulants
and internal emotions, that it is not easy to dis-
tinguish what portion of the tendency to such
alternations depends on original constitution. Yet
no one can doubt that the inclination to food and
sleep is periodical, or can maintain, with any

plausibility, that the period may be lengthened or shortened without limit. We may be tolerably certain that a constantly recurring period of forty-eight hours would be too long for one day of employment and one period of sleep, with our present faculties; and all, whose bodies and minds are tolerably active, will probably agree that, independently of habit, a perpetual alternation of eight hours up and four in bed would employ the human powers less advantageously and agreeably than an alternation of sixteen and eight. A creature which could employ the full energies of his body and mind uninterruptedly for nine months, and then take a single sleep of three months, would not be a man.

When, therefore, we have subtracted from the daily cycle of the employments of men and animals, that which is to be set down to the account of habits acquired, and that which is occasioned by extraneous causes, there still remains a periodical character; and a period of a certain length, which coincides with, or at any rate easily accommodates itself to, the duration of the earth's revolution. The physiological analysis of this part of our constitution is not necessary for our purpose. The succession of exertion and repose in the muscular system, of excited and dormant sensibility in the nervous, appear to be fundamentally connected with the muscular and nervous powers, whatever the nature of these may be. The necessity of these alternations is one of the measures of the intensity of those vital energies ; and it would seem that we cannot, without assuming the human powers to be altered, suppose the intervals of tranquillity which they require to be much changed. This view agrees with the opinion of some of the most eminent physiologists. Thus Cabanis* notices the periodical and isochronous character of the desire of sleep, as well as of other appetites. He states also that sleep is more easy and more salutary, in proportion as we go to rest and rise every day at the same hours ; and observes that this periodicity seems to have a reference to the motions of the solar system.

Now how should such a reference be at first established in the constitution of man, animals, and plants, and transmitted from one generation of them to another? If we suppose a wise and benevolent Creator, by whom all the parts of nature were fitted to their uses and to each other, this is what we might expect and can understand. On any other supposition such a fact appears altogether incredible and inconceivable.

* Rapports du Physique et du Moral de l'Homme, II. 371.

Chapter III.

The Mass of the Earth.

We shall now consider the adaptation which may, as we conceive, be traced in the amount of some of the quantities which determine the course of events in the organic world ; and especially in the amount of the *forces* which are in action. The life of vegetables and animals implies a constant motion of their fluid parts, and this motion must be produced by forces which urge or draw the particles of the fluids. The positions of the parts of vegetables are also the result of the flexibility and elasticity of their substance ; the voluntary motions of animals are produced by the tension of the muscles. But in all those cases, the effect really produced depends upon the force of gravity also ; and in order that the motions and positions may be such as answer their purpose, the forces which produce them must have a due proportion to the force of gravity. In human works, if, for instance, we have a fluid to raise, or a weight to move, some calculation is requisite, in order to determine the power which we must use, relatively to the work which is to be done : we have a mechanical problem to solve, in order that we may adjust the one to the other. And the same adjustment, the same result of a comparison of quantities, manifests itself in the relation which the forces of the organic world bear to the force of gravity.

The force of gravity might, so far as we can judge, have been different from what it now is. It depends upon the mass of the earth; and this mass is one of the elements of the solar system, which is not determined by any cosmical necessity of which we are aware. The masses of the several planets are very different, and do not appear to follow any determinate rule, except that upon the whole those nearer to the sun appear to be smaller, and those nearer the outskirts of the system to be larger. We cannot see anything which would have prevented either the size or the density of the earth from being different, to a very great extent, from what they are.

Now, it will be very obvious that if the intensity of gravity were to be much increased, or much diminished, if every object were to become twice as heavy or only half as heavy as it now is, all the forces, both of involuntary and voluntary motion which produce the present orderly and suitable results by being properly proportioned to the resistance which they experience, would be thrown off their balance; they would produce motions too quick or too slow, wrong positions, jerks and stops, instead of steady, well conducted movements. The universe would be like a machine ill regulated; every thing would go wrong; repeated collisions and a rapid disorganization must be the consequence. We will, however, attempt to illustrate one or two of the cases in which this would take place, by pointing out forces which act in the organic world, and which are adjusted to the force of gravity.

1. The first instance we shall take, is the force manifested by the ascent of the sap in vegetables. It appears, by a multitude of indisputable experiments, (among the rest, those of Hales, Mirbel, and Dutrochet,) that all plants imbibe moisture by their roots, and *pump it up*, by some internal force, into every part of their frame, distributing it into every leaf. It will easily be conceived that this operation must require a very considerable mechanical force; for the fluid must be sustained as if it were a single column reaching to the top of the tree. The division into minute parts and distribution through small vessels does not at all diminish the total force requisite to raise it. If, for instance, the tree be thirty-three feet high, the pressure must be fifteen pounds upon every square inch in the section of the vessels of the bottom, in order merely to support the sap. And it is not only supported, but propelled upwards with great force, so as to supply the constant evaporation of the leaves. The pumping power of the tree must, therefore, be very considerable.

That this power is great, has been confirmed by various curious experiments, especially by those of Hales. He measured the force with which the stems and branches of trees draw the fluid from below, and push it upwards. He found, for instance, that a vine in the *bleeding* season could push up its sap in a glass tube to the height of twenty-one feet above the stump of an amputated branch.

The force which produces this effect is part of the economy of the vegetable world; and it is clear that the due operation of the force depends upon its being rightly proportioned to the force of gravity. The weight of the fluid must be counterbalanced, and an excess of force must exist to produce the motion upwards. In the common course of vegetable life, the rate of ascent of the sap is regulated, on the one hand, by the upward pressure of the vegetable power, and on the other, by the amount of the gravity of the fluid, along with the other resistances, which are to be overcome. If, therefore, we suppose gravity to increase, the rapidity of this vegetable circulation will diminish, and the rate at which this function proceeds, will not correspond either to the course of the seasons, or the other physiological processes with which this has to co-operate. We might easily conceive such an increase of gravity as would stop the vital movements of the plant in a very short time. In like manner, a diminution of the gravity of the vegetable juices would accelerate the rising of the sap, and would, probably, hurry and overload the leaves and other organs, so as to interfere with their due operation. Some injurious change, at least, would take place.

Here, then, we have the forces of the minutest parts of vegetables adjusted to the magnitude of the whole mass of the earth on which they exist. There is no apparent connexion between the quantity of matter of the earth, and the force of imbibition of the roots of a vine, or the force of propulsion of the vessels of its branches. Yet, these things have such a proportion as the well-being of the vine requires. How is this to be accounted for, but by supposing that the circumstances under which the vine was to grow, were attended to in devising its structure?

We have not here pretended to decide whether this force of propulsion of vegetables is mechanical or not, because the argument is the same for our purpose on either supposition. Some very curious experiments have recently been made, (by M. Dutrochet) which are supposed to shew that the force is mechanical; that when two different fluids are separated by a thin membrane, a force which M. Dutrochet calls *endosmose* urges one fluid through the membrane: and that the roots of plants are provided with small vesicles which act the part of such a membrane. M. Poisson has further attempted to shew that this force of *endos-mose* may be considered as a particular modification of capillary action. If these views be true, we have here two mechanical forces, capillary action and gravity, which are adjusted to each other in the manner precisely suited to the welfare of vegetables.

2. As another instance of adaptation between the force of gravity and forces which exist in the vegetable world, we may take the positions of flowers. Some flowers grow with the hollow of their cup upwards : others "hang the pensive head" and turn the opening downwards. Now of these "nodding flowers" as Linnæus calls them, he observes that they are such as have their pistil longer than the stamens ; and, in consequence of this position, the dust from the anthers which are at the ends of the stamens can fall upon the stigma or extremity of the pistil ; which process is requisite for making the flower fertile. He gives as instances the flowers *campanula, leucoium, galanthus, fritillaria.* Other botanists have remarked that the position changes at different periods of the flower's progress. The pistil of the Euphorbia (which is a little globe or germen on a slender stalk) grows upright at first, and is taller than the stamens : at the period suited to its fecundation, the stalk bends under the weight of the ball at its extremity, so as to depress the germen below the stamens : after this

it again becomes erect, the globe being now a fruit filled with fertile seeds.

The positions in all these cases depend upon the length and flexibility of the stalk which supports the flower, or in the case of the Euphorbia, the germen. It is clear that a very slight alteration in the force of gravity, or in the stiffness of the stalk, would entirely alter the position of the flower cup, and thus make the continuation of the species impossible. We have therefore here a little mechanical contrivance, which would have been frustrated if the proper intensity of gravity had not been assumed in the reckoning. An earth greater or smaller, denser or rarer than the one on which we live, would require a change in the structure and strength of the footstalks of all the little flowers that hang their heads under our hedges. There is something curious in thus considering the whole mass of the earth from pole to pole, and from circumference to centre, as employed in keeping a snowdrop in the position most suited to the promotion of its vegetable health.

It would be easy to mention many other parts of the economy of vegetable life, which depend for their use on their adaptation to the force of gravity. Such are the forces and conditions which determine the positions of leaves and of branches. Such again those parts of the vegetable constitution which have reference to the pressure of the atmosphere ; for differences in this pressure appear to exercise a powerful influence on the functions of plants, and to require differences of structure. But we pass over these considerations. The slightest attention to the relations of natural objects will shew that the subject is inexhaustible ; and all that we can or need do is to give a few examples, such as may shew the nature of the impression which the examination of the universe produces.

3. Another instance of the adjustment of organic structure to the force of gravity may be pointed out in the muscular powers of animals. If the force of gravity were increased in any considerable proportion at the surface of the earth, it is manifest that all the swiftness, and strength, and grace of animal motions must disappear If, for instance, the earth were as large as Jupiter, gravity would be eleven times what it is, the lightness of the fawn, the speed of the hare, the spring of the tiger, could no longer exist with the existing muscular powers of those animals ; for man to lift himself upright, or to crawl from place to place, would be a labour slower and more painful than the motions of the sloth. The density and pressure of the air too would be increased to an intolerable extent, and the operation of respiration, and others, which depend upon these mechanical properties, would be rendered laborious, ineffectual, and probably impossible.

If, on the other hand, the force of gravity were much lessened, inconveniences of an opposite kind would occur. The air would be too thin to breathe ; the weight of our bodies, and of all the substances surrounding us, would become too slight to resist the perpetually occurring causes of derangement and unsteadiness : we should feel a want of ballast in our movements.

It has sometimes been maintained by fanciful theorists that the earth is merely a shell, and that the central parts are hollow. All the reasons we can collect appear to be in favour of its being a solid mass, considerably denser than any known rock. If this be so, and if we suppose the interior to be at any time scooped out, so as to leave only such a shell as the above mentioned speculators have asserted, we should not be left in ignorance of the change, though the appearance of the surface might remain the same. We should discover the want of the usual force of gravity, by the instability of all about us. Things would not lie where we placed them, but would slide away with the slightest push. We should have a difficulty in standing or walking, something like what we have on ship-board when the deck is inclined ; and we should stagger helplessly through an atmosphere thinner than that which oppresses the respiration of the traveller on the tops of the highest mountains.

We see therefore that those dark and unknown

central portions of the earth, which are placed far beyond the reach of the miner and the geologist, and of which man will probably never know anything directly, are not to be considered as quite disconnected with us, as deposits of useless lumber without effect or purpose. We feel their influence on every step we take and on every breath we draw; and the powers we possess, and the comforts we enjoy would be unprofitable to us, if they had not been prepared with a reference to those as well as to the near and visible portions of the earth's mass.

The arbitrary quantity, therefore, of which we have been treating, the intensity of the force of gravity, appears to have been taken account of, in establishing the laws of those forces by which the processes of vegetable and animal life are carried on. And this leads us inevitably, we conceive, to the belief of a supreme contriving mind, by which these laws were thus devised and thus established.

CHAPTER IV.

The Magnitude of the Ocean.

THERE are several arbitrary quantities which contribute to determine the state of things at the earth's surface besides those already mentioned. Some of these we shall briefly refer to, without pursuing the subject into detail. We wish not only to shew that the properties and processes of vegetable and animal life must be adjusted to each of these quantities in particular, but also to point out how numerous and complicated the conditions of the existence of organized beings are ; and we shall thus be led to think less inadequately of the intelligence which has embraced at once, and combined without confusion, all these conditions. We appear thus to be conducted to the conviction not only of design and intention, but of supreme knowledge and wisdom.

One of the quantities which enters into the constitution of the terrestrial system of things is the bulk of the waters of the ocean. The mean depth of the sea, according to the calculations of Laplace, is four or five miles. On this supposition, the addition to the sea of one-fourth of the existing waters would drown the whole of the globe, except a few chains of mountains. Whether this be exact or no, we can easily conceive the quantity of water which lies in the cavities of our globe to be greater or less than it at present is. With every such addition or subtraction the form and magnitude of the dry land would vary, and if this change were considerable, many of the present relations of things would be altered. It may be sufficient to mention one effect of such a change. The sources which water the earth, both clouds, rains, and rivers, are mainly fed by the aqueous vapour raised from the sea ; and therefore if the sea were much diminished, and the land increased, the mean quantity of moisture distributed upon the land must be diminished, and the character of climates, as to wet and dry, must be materially affected. Similar, but opposite changes would result from the increase of the surface of the ocean.

It appears then that the magnitude of the ocean is one of the conditions to which the structure of all organized beings which are dependent upon climate must be adapted.

CHAPTER V.

The Magnitude of the Atmosphere.

THE total quantity of air of which our atmosphere is composed is another of the arbitrary magnitudes of our terrestrial system ; and we may apply to this subject considerations similar to those of the last section. We can see no reason why the atmosphere might not have been larger in comparison to the globe which it surrounds ; those of Mars and Jupiter appear to be so. But if the quantity of air were increased, the structure of organized beings would in many ways cease to be adapted to their place. The atmospheric pressure, for instance, would be increased, which, as we have already noticed, would require an alteration in the structure of vegetables.

Another way in which an increase of the mass of the atmosphere would produce inconvenience would be in the force of winds. If the current of air in a strong gale were doubled or tripled, as might be the case if the atmosphere were augmented, the destructive effects would be more than doubled or tripled. With such a change, nothing could stand against a storm. In general, houses and trees resist the violence of the wind ; and except in extreme cases, as for instance in occasional hurricanes in the West Indies, a few large trees in a forest are unusual trophies of the power of the tempest. The breezes which we commonly have are harmless messengers to bring about the salutary changes of the atmosphere, even the motion which they communicate to vegetables tends to promote their growth, and is so advantageous, that it has been proposed to imitate it by artificial breezes in the hothouse. But with a stream of wind blowing against them, like three, or five, or ten, gales compressed into the space of one, none of the existing trees could stand; and except they could either bend like rushes in a stream, or extend their roots far wider than their branches, they must be torn up in whole groves. We have thus a manifest adaptation of the present usual strength of the materials and of the workmanship of the world to the stress of wind and weather which they have to sustain.

Chapter VI.

The Constancy and Variety of Climates.

IT is possible to conceive arrangements of our system, according to which all parts of the earth might have the same, or nearly the same, climate. If, for example, we suppose the earth to be a flat disk, or flat ring, like the ring of Saturn, revolving in its own plane as that does, each part of both the flat surfaces would have the same exposure to the sun, and the same temperature, so far as the sun's effect is concerned. There is no obvious reason why a planet of such a form might not be occupied by animals and vegetables, as well as our present earth; and on this supposition the climate would be every where the same, and the whole surface might be covered with life, without the necessity of there being any difference in the kind of inhabitants belonging to different parts.

Again, it is possible to conceive arrangements according to which no part of our planet should have any steady climate. This may probably be the case with a comet. If we suppose such a body, revolving round the sun in a very oblong ellipse, to be of small size and of a very high temperature, and therefore to cool rapidly; and if we suppose it also to be surrounded by a large atmosphere, composed of various gases; there would, on the surface of such a body, be no average climate or seasons for each place. The years, if we give this name to the intervals of time occupied by its successive revolutions, would be entirely unlike one another. The greatest heat of one year might be cool compared with the greatest cold of a preceding one. The greatest heats and colds might succeed each other at intervals perpetually unequal. The atmosphere might be perpetually changing its composition by the condensation of some of its constituent gases. In the operations of the elements, all would be incessant and rapid change, without recurrence or compensation. We cannot say that organized beings could not be fitted for such a habitation ; but if they were, the adaptation must be made by means of a constitution quite different from that of almost all organized beings known to us.

The state of things upon the earth, in its present condition, is very different from both these suppositions. The climate of the same place, notwithstanding perpetual and apparently irregular change, possesses a remarkable steadiness. And, though in different places the annual succession of appearances in the earth and heavens, is, in some of its main characters, the same, the result of these influences in the average climate is very different.

Now, to this remarkable constitution of the earth as to climate, the constitution of the animal and vegetable world is precisely adapted. The differences of different climates are provided for by the existence of entirely different classes of plants and animals in different countries. The constancy of climate at the same place is a necessary condition of the prosperity of each species there fixed.

We shall illustrate, by a few details, these characteristics in the constitution of inorganic and of organic nature, with the view of fixing the reader's attention upon the correspondence of the two.

1. The succession and alternation, at any given place, of heat and cold, rain and sunshine, wind and calm, and other atmospheric changes, appears at first sight to be extremely irregular, and not subject to any law. It is, however, easy to see, with a little attention, that there is a certain degree of constancy in the average weather and seasons of each place, though the particular facts of which these generalities are made up seem to be out of the reach of fixed laws. And when we apply any numerical measure to these particular occurrences, and take the average of the numbers thus observed, we generally find a remarkably close correspondence in the numbers belonging to the whole, or to analogous portions of successive years. This will be found to apply to the measures given by the thermometer, the barometer, the hygrometer, the raingage, and similar instruments. Thus it is found that very hot summers, or very cold winters, raise or depress the mean annual temperature very little above or below the general standard.

The heat may be expressed by degrees of the thermometer; the temperature of the day is estimated by this measure taken at a certain period of the day, which is found by experience to correspond with the daily average; and the mean annual temperature will then be the average of all the heights of the thermometer for every day in the year.

The mean annual temperature of London, thus measured, is about 50 degrees 4-10ths. The frost of the year 1788 was so severe that the Thames was passable on the ice; the mean temperature of that year was 50 degrees 6-10ths, being within a small fraction of a degree of the standard. In 1796, when the greatest cold ever observed in London occurred, the mean temperature of the year was 50 degrees 1-10th, which is likewise within a fraction of a degree of the standard. In the severe winter of 1813-14, when the Thames, Tyne, and other large rivers in England were completely frozen over, the mean temperature of the two years was 49 degrees, being little more than a degree below the standard. And in the year 1808, when the summer was so hot that the temperature in London was as high as 93½ degrees, the mean heat of the year was 50½, which is about that of the standard.

The same numerical indications of the constancy of climate at the same place might be collected from the records of other instruments of the kind above mentioned.

We shall, hereafter, consider some of the very complex agencies by which this steadiness is produced; and shall endeavour to point out intentional adaptations to this object. But we may, in the meantime, observe how this property of the atmospheric changes is made subservient to a further object.

To this constancy of the climates of each place, the structure of plants is adapted; almost all vegetables require a particular mean temperature of the year, or of some season of the year; a particular degree of moisture, and similar conditions. This will be seen by observing that the range of most plants as to climate is very limited. A vegetable which flourishes where the mean temperature is 55 degrees, would pine and wither when removed to a region where the average is 50 degrees. If, therefore, the average at each place were to vary as much as this, our plants with their present constitutions would suffer, languish, and soon die.

2. It will be readily understood that the same mode of measurement by which we learn the constancy of climate at the same place, serves to shew us the variety which belongs to different places. While the variations of the same region vanish when we take the averages even of moderate periods, those of distant countries are fixed and perpetual; and stand out more clear and distinct, the longer is the interval for which we measure their operation.

In the way of measuring already described, the mean temperature of Petersburg is 39 degrees, of Rome 60, of Cairo 72. Such observations as these, and others of the same kind, have been made at various places, collected and recorded ; and in this way the surface of the earth can be divided by boundary lines into various strips, according to these physical differences. Thus, the zones which take in all the places having the same or nearly the same mean annual temperature, have been called *isothermal* zones. These zones run nearly parallel to the equator, but not exactly, for, in Europe, they bend to the north in going eastward. In the same manner, the lines passing through all places which have an equal temperature for the summer or the winter half of the year, have been called respectively *isotheral* and *isochimal* lines. These do not coincide with the isothermal lines, for a place may have the same temperature as another, though its summer be hotter and its winter colder, as is the case of Pekin compared with London. In the same way we might conceive lines drawn according to the conditions of clouds, rain, wind, and the like circumstances, if we had observations enough to enable us to lay down such lines. The course of vegetation depends upon the combined influence of all such conditions ; and the lines which bound the spread of particular vegetable productions do not, in most cases, coincide with any of the separate meteorological boundaries above spoken of. Thus, the northern limit of vineyards runs through France, in a direction very nearly north-

east and south-west, while the line of equal temperature is nearly east and west. And the spon-
taneous growth or advantageous cultivation of
other plants, is in like manner bounded by lines
of which the course depends upon very complex
causes, but of which the position is generally
precise and fixed.

CHAPTER VII.

*The Variety of Organization corresponding to
the Variety of Climate.*

THE organization of plants and animals is in
different tribes formed upon schemes more or less
different, but in all cases adjusted in a general
way to the course and action of the elements.
The differences are connected with the different
habits and manners of living which belong to
different species ; and at any one place the
various species, both of animals and plants, have
a number of relations and mutual dependences
arising out of these differences. But besides the
differences of this kind, we find in the forms of
organic life another set of differences, by which
the animal and vegetable kingdom are fitted for
that variety in the climates of the earth, which
we have been endeavouring to explain.

The existence of such differences is too obvious
to require to be dwelt upon. The plants and

animals which flourish and thrive in countries remote from each other, offer to the eye of the traveller a series of pictures, which, even to an ignorant
and unreflective spectator, is full of a peculiar
and fascinating interest in consequence of the
novelty and strangeness of the successive scenes

Those who describe the countries between the
tropics, speak with admiration of the luxuriant
profusion and rich variety of the vegetable productions of those regions. Vegetable life seems
there far more vigorous and active, the circumstances under which it goes on, far more favourable
than in our latitudes. Now if we conceive an
inhabitant of those regions, knowing, from the
circumstances of the earth's form and motion, the
difference of climates which must prevail upon it,
to guess, from what he saw about him, the condition of other parts of the globe as to vegetable
wealth, is it not likely that he would suppose that
the extratropical climates must be almost devoid
of plants? We know that the ancients, living in
the temperate zone, came to the conclusion that
both the torrid and the frigid zones must be
uninhabitable. In like manner the equatorial
reasoner would probably conceive that vegetation
must cease, or gradually die away, as he should
proceed to places further and further removed
from the genial influence of the sun. The mean
temperature of his year being about 80 degrees,
he would hardly suppose that any plants could

subsist through a year, where the mean temperature was only 50, where the temperature of the
summer quarter was only 64, and where the
mean temperature of a whole quarter of the year
was a very few degrees removed from that at
which water becomes solid. He would suppose
that scarcely any tree, shrub, or flower could
exist in such a state of things, and so far as the
plants of his own country are concerned he would
judge rightly

But the countries further removed from the
equator are not left thus unprovided. Instead of
being scantily occupied by such of the tropical
plants as could support a stunted and precarious
life in ungenial climes, they are abundantly
stocked with a multitude of vegetables which
appear to be constructed expressly for them, inasmuch as these species can no more flourish at
the equator than the equatorial species can in
these temperate regions. And such new supplies
thus adapted to new conditions, recur perpetually
as we advance towards the apparently frozen and
untenantable regions in the neighbourhood of the
pole. Every zone has its peculiar vegetables ;
and as we miss some, we find others make their
appearance, as if to replace those which are
absent.

If we look at the indigenous plants of Asia and
Europe, we find such a succession as we have
here spoken of. At the equator we find the

natives of the Spice Islands, the clove and nutmeg trees, pepper and mace. Cinnamon bushes clothe the surface of Ceylon ; the odoriferous sandal wood, the ebony tree, the teak tree, the banyan, grow in the East Indies. In the same latitudes in Arabia the Happy we find balm, frankincense and myrrh, the coffee tree, and the tamarind. But in these countries, at least in the plains, the trees and shrubs which decorate our more northerly climes are wanting. And as we go northwards, at every step we change the vegetable group, both by addition and by subtraction. In the thickets to the west of the Caspian Sea we have the apricot, citron, peach, walnut. In the same latitude in Spain, Sicily, and Italy, we find the dwarf palm, the cypress, the chestnut, the cork tree : the orange and lemon tree perfume the air with their blossoms ; the myrtle and pomegranate grow wild among the rocks. We cross the Alps, and we find the vegetation which belongs to northern Europe, of which England is an instance. The oak, the beech, and the elm are natives of Great Britain : the elm tree seen in Scotland, and in the north of England, is the wych elm. As we travel still further to the north the forests again change their character. In the northern provinces of the Russian empire are found forests of the various species of firs : the scotch and spruce fir, and the larch. In the Orkney Islands no tree is found but the hazel, which occurs again on the

northern shores of the Baltic. As we proceed
into colder regions we still find species which
appear to have been made for these situations.
The hoary or cold elder makes its appearance
north of Stockholm : the sycamore and mountain
ash accompany us to the head of the gulf of
Bothnia : and as we leave this and traverse the
Dophrian range, we pass in succession the
boundary lines of the spruce fir, the scotch fir,
and those minute shrubs which botanists distin-
guish as the dwarf birch and dwarf willow.
Here, near to or within the arctic circle, we yet
find wild flowers of great beauty ; the mezereum,
the yellow and white water lily, and the European
globe flower. And when these fail us, the rein-
deer moss still makes the country habitable for
animals and man.

We have thus a variety in the laws of vegetable
organization remarkably adapted to the variety of
climates ; and by this adaptation the globe is
clothed with vegetation and peopled with animals
from pole to pole, while without such an adaptation vegetable and animal life must have been
confined almost, or entirely, to some narrow zone
on the earth's surface. We conceive that we see
here the evidence of a wise and benevolent intention, overcoming the varying difficulties, or employing the varying resources of the elements, with
an inexhaustible fertility of contrivance, a constant tendency to diffuse life and well being.

2. One of the great uses to which the vegetable wealth of the earth is applied, is the support of man, whom it provides with food and clothing ; and the adaptation of tribes of indigenous vegetables to every climate has, we cannot but believe, a reference to the intention that the human race should be diffused over the whole globe. But this end is not answered by indigenous vegetables alone ; and in the variety of vegetables capable of being *cultivated* with advantage in various countries, we conceive that we find evidence of an additional adaptation of the scheme of organic life to the system of the elements.

The cultivated vegetables, which form the necessaries or luxuries of human life, are each confined within limits, narrow, when compared with the whole surface of the earth ; yet almost every part of the earth's surface is capable of being abundantly covered with one kind or other of these. When one class fails, another appears in its place. Thus corn, wine, and oil, have each its boundaries. Wheat extends through the old Continent, from England to Thibet : but it stops soon in going northwards, and is not found to succeed in the west of Scotland. Nor does it thrive better in the torrid zone than in the polar regions: within the tropics, wheat, barley and oats are not cultivated, excepting in situations considerably above the level of the sea : the inhabitants of those countries have other species of grain, or

other food. The cultivation of the vine succeeds only in countries where the annual temperature is between 50 and 63 degrees. In both
hemispheres, the profitable culture of this plant
ceases within 30 degrees of the equator, unless in
elevated situations, or in islands, as Teneriffe.
The limits of the cultivation of maize and of olives
in France are parallel to those which bound the
vine and corn in succession to the north. In the
north of Italy, west of Milan, we first meet with
the cultivation of rice; which extends over all the
southern part of Asia, wherever the land can be
at pleasure covered with water. In great part of
Africa millet is one of the principal kinds of
grain.

Cotton is cultivated to latitude 40 in the new
world, but extends to Astrachan in latitude 46 in
the old. The sugar cane, the plantain, the mulberry, the betel nut, the indigo tree, the tea tree,
repay the labours of the cultivator in India and
China; and several of these plants have been
transferred, with success, to America and the
West Indies. In equinoctial America a great
number of inhabitants find abundant nourishment
on a narrow space cultivated with plantain,
cassava yams, and maize. The bread fruit tree
begins to be cultivated in the Manillas, and
extends through the Pacific; the sago palm in
the Moluccas, the cabbage tree in the Pelew
islands.

In this manner the various tribes of men are provided with vegetable food. Some however live on their cattle, and thus make the produce of the earth only mediately subservient to their wants. Thus the Tatar tribes depend on their flocks and herds for food : the taste for the flesh of the horse seems to belong to the Mongols, Fins, and other descendants of the ancient Scythians : the locust eaters are found now, as formerly, in Africa.

Many of these differences depend upon custom, soil, and other causes with which we do not here meddle ; but many are connected with climate : and the variety of the resources which man thus possesses, arises from the variety of constitution belonging to cultivable vegetables, through which one is fitted to one range of climate, and another to another. We conceive that this variety and succession of fitness for cultivation, shews undoubted marks of a most foreseeing and benevolent design in the Creator of man and of the world.

3. By differences in vegetables of the kind we have above described, the sustentation and gratification of man's physical nature is copiously provided for. But there is another circumstance, a result of the difference of the native products of different regions, and therefore a consequence of that difference of climate on which the difference of native products depends,* which appears to be

* It will be observed that it is not here asserted that the difference of native products depends on the difference of climate *alone*.

worthy our notice. The difference of the productions of different countries has a bearing not
only upon the physical, but upon the social and
moral condition of man.

The intercourse of nations in the way of
discovery, colonization, commerce; the study of
the natural history, manners, institutions of foreign
countries; lead to most numerous and important
results. Without dwelling upon this subject, it will
probably be allowed that such intercourse has a
great influence upon the comforts, the prosperity,
the arts, the literature, the power, of the nations
which thus communicate. Now the variety of the
productions of different lands supplies both the
stimulus to this intercourse, and the instruments
by which it produces its effects. The desire to
possess the objects or the knowledge which
foreign countries alone can supply, urges the
trader, the traveller, the discoverer to compass
land and sea; and the progress of the arts and
advantages of civilization consists almost entirely
in the cultivation, the use, the improvement of that
which has been received from other countries.

This is the case to a much greater extent than
might at first sight be supposed. Where man
is active as a cultivator, he scarcely ever bestows
much of his care on those vegetables which the
land would produce in a state of nature. He
does not select some of the plants of the soil and
improve them by careful culture, but, for the most

part, he expels the native possessors of the land, and introduces colonies of strangers.

Thus, to take the condition of our own part of the globe as an example; scarcely one of the plants which occupy our fields and gardens is indigenous to the country. The walnut and the peach come to us from Persia; the apricot from Armenia : from Asia Minor, and Syria, we have the cherry tree, the fig, the pear, the pome-granate, the olive, the plum, and the mulberry The vine which is now cultivated is not a native of Europe ; it is found wild on the shores of the Caspian, in Armenia and Caramania. The most useful species of plants, the *cereal* vegetables, are certainly strangers, though their birth place seems to be an impenetrable secret. Some have fancied that barley is found wild on the banks of the Semara, in Tartary, rye in Crete, wheat at Baschkiros, in Asia ; but this is held by the best botanists to be very doubtful. The potatoe, which has been so widely diffused over the world in modern times, and has added so much to the resources of life in many countries, has been found equally difficult to trace back to its wild con-dition.

Thus widely are spread the traces of the con-nexion of the progress of civilization with na-tional intercourse. In our own country a higher state of the arts of life is marked by a more ready and extensive adoption of foreign produc-

tions. Our fields are covered with herbs from
Holland, and roots from Germany ; with Flemish
farming and Swedish turnips ; our hills with
forests of the firs of Norway. The chestnut and
poplar of the south of Europe adorn our lawns,
and below them flourish shrubs and flowers from
every clime in profusion. In the mean time
Arabia improves our horses, China our pigs,
North America our poultry, Spain our sheep,
and almost every country sends its dog. The
products which are ingredients in our luxuries,
and which we cannot naturalize at home, we
raise in our colonies ; the cotton, coffee, sugar of
the east are thus transplanted to the farthest west;
and man lives in the middle of a rich and varied
abundance which depends on the facility with
which plants and animals and modes of culture
can be transferred into lands far removed from
those in which nature had placed them. And
this plenty and variety of material comforts is
the companion and the mark of advantages and
improvements in social life, of progress in art
and science, of activity of thought, of energy of
purpose, and of ascendency of character.

The differences in the productions of different
countries which lead to the habitual intercourse
of nations, and through this to the benefits which
we have thus briefly noticed, do not all depend
upon the differences of temperature and climate
alone. But these differences are among the

causes, and are some of the most important causes, or conditions, of the variety of products; and thus that arrangement of the earth's form and motion from which the different climates of different places arises, is connected with the social and moral welfare and advancement of man.

We conceive that this connexion, though there must be to our apprehension much that is indefinite and uncertain in tracing its details, is yet a point where we may perceive the profound and comprehensive relations established by the counsel and foresight of a wise and good Creator of the world and of man, by whom the progress and elevation of the human species was neither uncontemplated nor uncared for.

4. We have traced, in the variety of organized beings, an *adaptation* to the variety of climates, a *provision* for the sustentation of man all over the globe, and an *instrument* for the promotion of civilization and many attendant benefits. We have not considered this *variety* as *itself* a purpose which we can perceive or understand without reference to some ulterior end. Many persons, however, and especially those who are already in the habit of referring the world to its Creator, will probably see something admirable in itself in this vast variety of created things. There is indeed something well fitted to produce and confirm a reverential wonder, in these apparently inexhaustible stores of new forms of being and

modes of existence ; the fixity of the laws of each
class, its distinctness from all others, its relations
to many. Structures and habits and characters are
exhibited, which are connected and distinguished
according to every conceivable degree of subordination and analogy, in their resemblances and in
their differences. Every new country we explore
presents us with new combinations, where the
possible cases seemed to be exhausted ; and with
new resemblances and differences, constructed
as if to elude what conjecture might have hit
upon, by proceeding from the old ones. Most of
those who have any large portion of nature
brought under their notice in this point of view,
are led to feel that there is, in such a creation, a
harmony, a beauty, and a dignity, of which the
impression is irresistible ; which would have been
wanting in any more uniform and limited system
such as we might try to imagine ; and which of
itself gives to the arrangements by which such
a variety on the earth's surface is produced, the
character of well devised means to a worthy end.

Chapter VIII.

The Constituents of Climate.

WE have spoken of the steady average of the climate at each place, of the difference of this average at different places, and of the adaptation of organized beings to this character in the laws of the elements by which they are affected. But this steadiness in the general effect of the elements, is the result of an extremely complex and extensive machinery. Climate, in its wider sense, is not one single agent, but is the aggregate result of a great number of different agents, governed by different laws, producing effects of various kinds. The steadiness of this compound agency is not the steadiness of a permanent condition, like that of a body at rest ; but it is the steadiness of a state of constant change and movement, succession and alternation, seeming accident and irregularity. It is a perpetual repose, combined with a perpetual motion ; an invariable average of most variable quantities. Now, the manner in which such a state of things is produced, deserves, we conceive, a closer consideration. It may be useful to shew how the particular laws of the action of each of the elements of climate are so adjusted that they do not disturb this general constancy.

The principal constituents of climate are the following :—the temperature of the earth, of the water, of the air :—the distribution of the aqueous vapour contained in the atmosphere :—the winds and rains by which the equilibrium of the atmosphere is restored when it is in any degree disturbed. The effects of light, of electricity, probably of other causes also, are no doubt important in the economy of the vegetable world, but these agencies have not been reduced by scientific inquirers to such laws as to admit of their being treated with the same exactness and certainty which we can obtain in the case of those first mentioned.

We shall proceed to trace some of the peculiarities in the laws of the different physical agents which are in action at the earth's surface, and the manner in which these peculiarities bear upon the general result.

The Laws of Heat with respect to the Earth.

One of the main causes which determine the temperature of each climate is the effect of the sun s rays on the solid mass of the earth. The laws of this operation have been recently made out with considerable exactness, experimentally by Leslie, theoretically by Fourrier, and by other inquirers. The theoretical inquiries have required the application of very complex and abstruse mathematical investigations ; but the general character of the operation may, perhaps, be made easily intelligible.

The earth, like all solid bodies, transmits into its interior the impressions of heat which it receives at the surface ; and throws off the superfluous heat from its surface into the surrounding space.   These processes are called *conduction* and *radiation*, and have each their ascertained mathematical laws.

By the laws of conduction, the daily impressions of heat which the earth receives, follow each other into the interior of the mass, like the waves which start from the edge of a canal* ; and like them, become more and more faint as they proceed, till they melt into the general level of the internal temperature.   The heat thus transmitted is accumulated in the interior of the earth, as in a reservoir, and flows from one part to another of this reservoir   The parts of the earth near the equator are more heated by the sun than other parts, and on this account there is a perpetual internal conduction of heat from the equatorial to other parts of the sphere.   And as all parts of the surface throw off heat by radiation, in the polar regions, where the surface receives little in return from the sun, a constant waste is produced.   There is thus from the polar parts a perpetual dispersion of heat in the surrounding space, which is supplied by a perpetual internal flow from the equator towards each pole.

---

* The resemblance consists in this; that we have a strip of greater temperature accompanied by a strip of smaller temperature, these strips arising from the diurnal and nocturnal impressions respectively, and being in motion; as in the waves on a canal, we have a moving strip of greater elevation accompanied by a strip of smaller elevation. We do not here refer to any hypothetical undulations in the fluid matter of heat.

Here, then, is a kind of circulation of heat ; and the quantity and rapidity of this circulation, determine the quantity of heat in the solid part of the earth, and in each portion of it ; and through this, the *mean* temperature belonging to each point on its surface.

If the earth *conducted* heat more rapidly than it does, the inequalities of temperature would be more quickly balanced, and the temperature of the ground (below the reach of annual and diurnal variations) would differ less than it does.   If the surface *radiated* more rapidly than it does, the flow of heat from the polar regions would increase, and the temperature of the interior of the globe would find a lower level ; the differences of temperature in different latitudes would increase, but the mean temperature of the globe would diminish.

There is nothing which, so far as we can perceive, determines necessarily, either the conducting or the radiating power of the earth to its present value.   The measures of such powers, in different substances, differ very widely.   If the earth were a globe of pure iron, it would conduct heat, probably, twenty times as well as it does; if its surface were polished iron, it would only radiate one-sixth as much as it does.   Changes in the amount of the conduction and radiation far less than these, would, probably, subvert the whole *thermal* constitution of the earth, and make it uninhabitable by any of its present vegetable, or animal tenants.

One of the results of the laws of heat, as they

exist in the globe, is, that, by their action, the thermal state tends to a limiting condition, which, once reached, remains constant and steady, as it now is. The oscillations or excursions from the mean condition, produced by any temporary cause, are rapidly suppressed ; the deviations of seasons from their usual standard produce only a small and transient effect. The impression of an extremely hot day upon the ground melts almost immediately into the average internal heat. The effect of a hot summer, in like manner, is soon lost in its progress through the globe. If this were otherwise, if the inequalities and oscillations of heat went on, through the interior of the earth, retaining the same value, or becoming larger and larger, we might have the extreme heats or colds of one place making their appearance at another place after a long interval ; like a conflagration which creeps along a street and bursts out at a point remote from its origin.

It appears, therefore, that both the present differences of climate, and the steadiness of the average at each place, depend upon the form of the present laws of heat, and on the arbitrary magnitudes which determine the rate of conduction and radiation. The laws are such as to secure us from increasing and destructive inequalities of heat ; the arbitrary magnitudes are elements to which the organic world is adjusted.

Chapter IX.

The Laws of Heat with respect to Water.

The manner in which heat is transmitted through fluids is altogether different from the mode in which it passes through solids; and hence the waters of the earth's surface produce peculiar effects upon its condition as to temperature. Moreover, water is susceptible of evaporation in a degree depending upon the increase of heat; and in consequence of this property it has most extensive and important functions to discharge in the economy of nature. We will consider some of the offices of this fluid.

1. Heat is communicated through water, not by being *conducted* from one part of the fluid to another, as in solid bodies, but (at least principally) by being *carried* with the parts of the fluid by means of an intestine motion. Water expands and becomes lighter by heat, and, therefore, if the upper parts be cooled below the subjacent temperature, this upper portion will become heavier than that below, bulk for bulk, and will descend through it, while the lower portion rises to take the upper place. In this manner the colder parts descend, and the warmer parts ascend by contrary currents, and by their interchange and

mixture, reduce the whole to a temperature at least as low as that of the surface. And this equalization of temperature by means of such currents, is an operation of a much more rapid nature than the slow motion of conduction by which heat creeps through a solid body. Hence, alternations of heat and cold, as day and night, summer and winter, produce in water, inequalities of temperature much smaller than those which occur in a solid body. The heat communicated is less, for transparent fluids imbibe heat very slowly; and the cold impressed on the surface is soon diffused through the mass by internal circulation.

Hence it follows that the ocean, which covers so large a portion of the earth, and affects the temperature of the whole surface by its influence, produces the effect of making the alternations of heat and cold much less violent than they would be if it were absent. The different temperatures of its upper and lower parts produce a current which draws the seas, and by means of the seas, the air, towards the mean temperature. And this kind of circulation is produced, not only between the upper and lower parts, but also between distant tracts of the ocean. The great Gulf Stream which rushes out of the gulf of Mexico, and runs across the Atlantic to the western shores of Europe, carries with it a portion of the tropical heat into northern regions: and the returning current which descends along the coast of Africa, tends to cool the parts nearer the equator. Great as the difference of temperature is in different climates, it would be still greater if there were not this equalizing and moderating power exerted constantly over the whole surface. Without this influence, it is probable that the two polar portions of the earth, which are locked in perpetual ice and snow, and almost destitute of life, would be much increased.

We find an illustration of this effect of the ocean on temperature, in the peculiarities of the climates of maritime tracts and islands. The climate of such portions of the earth, corrected in some measure by the temperature of the neighbouring sea, is more equable than that of places in the same latitudes differently situated. London is cooler in summer and warmer in winter than Paris.

2. Water expands by heat and contracts by cold, as has been already said; and in consequence of this property, the coldest portions of the fluid generally occupy the lower parts. The continued progress of cold produces congelation. If, therefore, the law just mentioned had been strictly true, the lower parts of water would have been first frozen; and being once frozen, hardly any heat applied at the surface could have melted them, for the warm fluid could not have descended through the colder parts. This is so far the case, that in a vessel containing ice at the bottom and

water at the top, Rumford made the upper fluid boil without thawing the congealed cake below.

Now, a law of water with respect to heat operating in this manner, would have been very inconvenient if it had obtained in our lakes and seas. They would all have had a bed of ice, increasing with every occasion, till the whole was frozen. We could have had no bodies of water, except such pools on the surfaces of these icy reservoirs as the summer sun could thaw, to be again frozen to the bottom with the first frosty night. The law of the regular contraction of water by cold till it became ice, would, therefore, be destructive of all the utility of our seas and lakes. How is this inconvenience obviated?

It is obviated by a modification of the law which takes place when the temperature approaches this limit. Water contracts by the increase of cold, till we come *near* the freezing temperature; but then, by a further increase of cold, it contracts no more, but expands till the point at which it becomes ice. It contracts in cooling down to 40 degrees of Fahrenheit's thermometer; in cooling further it expands, and when cooled to 32 degrees, it freezes. Hence, the greatest density of the fluid is at 40 degrees, and water of this temperature, or near it, will lie at the bottom with cooler water or with ice floating above it. However much the surface be cooled, water colder than 40 cannot descend to displace water warmer than itself   Hence we can never have ice formed at the bottom of deep water. In approaching the freezing point, the coldest water will rise to the surface, and the congelation will take place there ; and the ice so formed will remain at the surface, exposed to the warmth of the sun-beams and the air, and will not survive any long continuance of such action.

Another peculiarity in the laws which regulate the action of cold on water is, that in the very act of freezing a further sudden and considerable expansion takes place.   Many persons will have known instances of vessels burst by the freezing of water in them.   The consequence of this expansion is, that the specific gravity of ice is less than that of water of any temperature ; and it therefore always floats in the unfrozen fluid.   If this expansion of crystallization did not exist, ice would float in water which was below 40 degrees, but would sink when the fluid was above that temperature : as the case is, it floats under all circumstances.   The icy remnants of the effects of winter, which the river carries down its stream, are visible on its surface till they melt away ; and the icebergs which are detached from the shores of the polar seas, drift along, exposed to the sun and air, as well as to the water in which they are immersed.

These laws of the effect of temperature on water are truly remarkable in their adaptation to the beneficial course of things at the earth's surface. Water contracts by cold; it thus equalizes the temperature of various times and places; but if its contraction were continued all the way to the freezing point, it would bind a great part of the earth in fetters of ice. The contraction then is here replaced by expansion, in a manner which but slightly modifies the former effects, while it completely obviates the bad consequences. The further expansion which takes place at the point of freezing, still further facilitates the rapid removal of the icy chains, in which parts of the earth's surface are at certain seasons bound.

We do not know how far these laws of expansion are connected with and depend on more remote and general properties of this fluid, or of all fluids. But we have no reason to believe that, by whatever means they operate, they are not laws selected from among other laws which might exist, as in fact for other fluids other laws do exist. And we have all the evidence, which the most remarkable furtherance of important purposes can give us, that they *are* selected, and selected with a beneficial design.

3. As water becomes ice by cold, it becomes steam by heat. In common language, steam is the name given to the vapour of *hot* water; but in fact a vapour or steam rises from water at all temperatures, however low, and even from ice. The

expansive force of this vapour increases rapidly as the heat increases; so that when we reach the heat of boiling water, it operates in a far more striking manner than when it is colder; but in all cases the surface of water is covered with an atmosphere of aqueous vapour, the pressure or *tension* of which is limited by the temperature of the water. To each degree of pressure in steam there is a *constituent temperature* corresponding. If the surface of water is not pressed by vapour with the force thus corresponding to its temperature, an immediate *evaporation* will supply the deficiency. We can compare the tension of such vapour with that of our common atmosphere; the pressure of the latter is measured by the barometrical column, about thirty inches of mercury; that of watery vapour is equal to one inch of mercury at the constituent temperature of 80 degrees, and to one-fifth of an inch, at the temperature of 32 degrees.

Hence, if that part of the atmosphere which consists of common air were annihilated, there would still remain an atmosphere of aqueous vapour, arising from the waters and moist parts of the earth; and in the existing state of things this vapour rises *in* the atmosphere of dry air. Its distribution and effects are materially influenced by the vehicle in which it is thus carried, as we shall hereafter notice; but at present we have to observe the exceeding *utility* of water in this shape. We remark how suitable and indispensable to the well-being of the creation it is, that the fluid should possess the property of assuming such a form under such circumstances.

The *moisture* which floats in the atmosphere is of most essential use to vegetable life.*   " The leaves of living plants appear to act upon this vapour in its elastic form, and to absorb it. Some vegetables increase in weight from this cause when suspended in the atmosphere and unconnected with the soil, as the house-leek and the aloe. In very intense heats, and when the soil is dry, the life of plants seems to be preserved by the absorbent power of their leaves."   It follows from what has already been said, that, with an increasing heat of the atmosphere, an increasing quantity of vapour will rise into it, if supplied from any quarter. Hence it appears that aqueous vapour is most abundant in the atmosphere when it is most needed for the purposes of life ; and that when other sources of moisture are cut off, this is most copious.

4. *Clouds* are produced by aqueous vapour when it returns to the state of water. This process is *condensation*, the reverse of evaporation. When vapour exists in the atmosphere, if in any manner the temperature becomes lower than the *constituent temperature*, requisite for the mainte-

* Loudon, 1219.

nance of the vapoury state, some of the steam will be condensed and will become water. It is in this manner that the curl of steam from the spout of a boiling tea-kettle becomes visible, being cooled down as it rushes to the air. The steam condenses into a fine watery powder, which is carried about by the little aerial currents. Clouds are of the same nature with such curls, the condensation being generally produced when air, charged with aqueous vapour, is mixed with a colder current, or has its temperature diminished in any other manner.

Clouds, while they retain that shape, are of the most essential use to vegetable and animal life. They moderate the fervour of the sun, in a manner agreeable, to a greater or less degree, in all climates, and grateful no less to vegetables than to animals. Duhamel says that plants grow more during a week of cloudy weather than a month of dry and hot. It has been observed that vegetables are far more refreshed by being watered in cloudy than in clear weather. In the latter case, probably the supply of fluid is too rapidly carried off by evaporation. Clouds also moderate the alternations of temperature, by checking the radiation from the earth. The coldest nights are those which occur under a cloudless winter sky.

The uses of clouds, therefore, in this stage of their history, are by no means inconsiderable,

and seem to indicate to us that the laws of their
formation were constructed with a view to the
purposes of organized life.

5. Clouds produce *rain*. In the formation of
a cloud the precipitation of moisture probably
forms a fine watery *powder*, which remains suspended in the air in consequence of the minuteness of its particles : but if from any cause the
precipitation is collected in larger portions, and
becomes *drops*, these descend by their weight
and produce a shower.

However rain is formed, it is one of the consequences of the capacity of evaporation and
condensation which belongs to water, and its
uses are the result of the laws of those processes.
Its uses to plants are too obvious and too numerous to be described. It is evident that on
its quantity and distribution depend in a great
measure the prosperity of the vegetable kingdom :
and different climates are fitted for different productions, no less by the relations of dry weather
and showers, than by those of hot and cold.

6. Returning back still further in the changes
which cold can produce on water, we come to
snow and *ice:* snow being apparently frozen
vapour, aggregated by a confused action of crystalline laws ; and ice being water in its fluid
state, solidified by the same crystalline forces.
The impression of these agents on the animal
feelings is generally unpleasant, and we are in

the habit of considering them as symptoms of the power of winter to interrupt that state of the elements in which they are subservient to life. Yet, even in this form, they are not without their uses.* " Snow and ice are bad conductors of cold ; and when the ground is covered with snow, or the surface of the soil or of water is frozen, the roots or bulbs of plants beneath are protected by the congealed water from the influence of the atmosphere, the temperature of which, in northern winters, is usually very much below the freezing point ; and this water becomes the first nourishment of the plant in early spring. The expansion of water during its congelation, at which time its volume increases one-twelfth, and its contraction in bulk during a thaw, tend to pulverize the soil, to separate its parts from each other, and to make it more permeable to the influence of the air." In consequence of the same slowness in the conduction of heat which snow thus possesses, the arctic traveller finds his bed of snow of no intolerable coldness ; the Esquimaux is sheltered from the inclemency of the season in his snow hut, and travels rapidly and agreeably over the frozen surface of the sea. The uses of those arrangements, which at first appear productive only of pain and inconvenience, are well suited to give

* Loudon, 1214.

confidence and hope to our researches for such usefulness in every part of the creation. They have thus a peculiar value in adding connexion and universality to our perception of beneficial design.

7. There is a peculiar circumstance still to be noticed in the changes from ice to water and from water to steam. These changes take place at a particular and invariable degree of heat; yet they do not take place suddenly when we increase the heat to this degree. This is a very curious arrangement. The temperature *makes a stand*, as it were, at the point where thaw, and where boiling take place. It is necessary to apply a considerable quantity of heat to produce these effects; all which heat disappears, or becomes *latent*, as it is called. We cannot raise the temperature of a thawing mass of ice till we have thawed the whole. We cannot raise the temperature of boiling water, or of steam rising from it, till we have converted all the water into steam. Any heat that we apply while these changes are going on is absorbed in producing the changes.

The consequences of this property of *latent heat* are very important. It is on this account that the changes now spoken of necessarily occupy a considerable time. Each part in succession must have a proper degree of heat applied to it. If it were otherwise, thaw and evaporation must be instantaneous: at the first touch of warmth, all the snow which lies on the roofs of our houses would descend like a waterspout into the streets : all that which rests on the ground would rush like an inundation into the water courses. The hut of the Esquimaux would vanish like a house in a pantomime : the icy floor of the river would be gone without giving any warning to the skaiter or the traveller : and when, in heating our water, we reached the boiling point, the whole fluid would " flash into steam," (to use the expression of engineers,) and dissipate itself in the atmosphere, or settle in dew on the neighbouring objects.

It is obviously necessary for the purposes of human life, that these changes should be of a more gradual and manageable kind than such as we have now described. Yet this gradual progress of freezing and thawing, of evaporation and condensation, is produced, so far as we can discover, by a particular contrivance. Like the freezing of water from the top, or the floating of ice, the moderation of the rate of these changes seems to be the result of a *violation* of a law : that is, the simple rule regarding the effects of change of temperature, which at first sight appears to be the law, and which, from its simplicity, would seem to us the most obvious law for these as well as other cases, is modified at certain critical points, *so as to* produce these advantageous effects :—why may we not say *in order to* produce such effects ?

8. Another office of water which it discharges by means of its relations to heat, is that of supplying our *springs*. There can be no doubt that the old hypotheses which represent springs as drawing their supplies from large subterranean reservoirs of water, or from the sea by a process of subterraneous filtration, are erroneous and untenable. The quantity of evaporation from water and from wet ground is found to be amply sufficient to supply the requisite drain. Mr. Dalton calculated* that the quantity of rain which falls in England is thirty-six inches a year. Of this he reckoned that thirteen inches flow off to the sea by the rivers, and that the remaining twenty-three inches are raised again from the ground by evaporation. The thirteen inches of water are of course supplied by evaporation from the sea, and are carried back to the land through the atmosphere. Vapour is perpetually rising from the ocean, and is condensed in the hills and high lands, and through their pores and crevices descends, till it is deflected, collected, and conducted out to the day, by some stratum or channel which is watertight. The condensation which takes place in the higher parts of a country, may easily be recognised in the mists and rains which are the frequent occupants of such regions. The coldness of the atmosphere and other causes precipitate the

* Manchester Memoirs, v. 357.

moisture in clouds and showers, and in the former
as well as in the latter shape, it is condensed and
absorbed by the cool ground. Thus a perpetual
and compound circulation of the waters is kept
up ; a narrower circle between the evaporation
and precipitation of the land itself, the rivers and
streams only occasionally and partially forming a
portion of the circuit ; and a wider interchange
between the sea and the lands which feed the
springs, the water ascending perpetually by a
thousand currents through the air, and descending
by the gradually converging branches of the rivers,
till it is again returned into the great reservoir of
the ocean.

In every country, these two portions of the
aqueous circulation have their regular, and nearly
constant, proportion. In this kingdom the relative quantities are, as we have said, 23 and 13.
A due distribution of these circulating fluids in
each country appears to be necessary to its
organic health ; to the habits of vegetables, and
of man. We have every reason to believe that
it is kept up from year to year as steadily as the
circulation of the blood in the veins and arteries
of man. It is maintained by a machinery very
different, indeed, from that of the human system,
but apparently as well, and, therefore, we may
say as clearly, as that, adapted to its purposes.

By this machinery, we have a connexion established between the atmospheric changes of

remote countries. Rains in England are often introduced by a south-east wind. " Vapour brought to us by such a wind, must have been generated in countries to the south and east of our island. It is therefore, probably, in the extensive valleys watered by the Meuse, the Moselle, and the Rhine, if not from the more distant Elbe, with the Oder and the Weser, that the water rises, in the midst of sunshine, which is soon afterwards to form *our* clouds, and pour down *our* thunder-showers." " Draught and sunshine in one part of Europe may be as necessary to the production of a wet season in another, as it is on the great scale of the continents of Africa and South America; where the plains, during one half the year, are burnt up, to feed the springs of the mountains ; which in their turn contribute to inundate the fertile valleys and prepare them for a luxuriant vegetation."* The properties of water which regard heat make one vast *watering-engine* of the atmosphere.

* Howard on the Climate of London, vol. ii. pp. 216, 217.

CHAPTER X.

The Laws of Heat with respect to Air.

WE have seen in the preceding chapter how many and how important are the offices discharged by the aqueous part of the atmosphere. The aqueous part is, however, a very small part only : it may vary, perhaps, from less than 1-100dth to nearly as much as 1-20th in weight, of the whole aerial ocean. We have to offer some considerations with regard to the remainder of the mass.

1. In the first place we may observe that the aerial atmosphere is necessary as a vehicle for the aqueous vapour. Salutary as is the operation of this last element to the whole organized creation, it is a substance which would not have answered its purposes if it had been administered pure. It requires to be diluted and associated with dry air, to make it serviceable. A little consideration will show this.

We can suppose the earth with no atmosphere except the vapour which arises from its watery parts : and if we suppose also the equatorial parts of the globe to be hot, and the polar parts cold, we may easily see what would be the consequence. The waters at the equator, and near the equator, would produce steam of greater elasticity, rarity, and temperature, than that which occupies the regions further *polewards;* and such steam, as it came in contact with the colder vapour of a higher latitude, would be precipitated into the form of water. Hence there would be a perpetual current of steam from the equatorial parts towards each pole, which would be condensed, would fall to the surface, and flow back to the equator in the form of fluid. We should have a circulation which might be regarded as a species of regulated distillation.* On a globe so constituted, the sky of the equatorial zone would be perpetually cloudless; but in all other latitudes we should have an uninterrupted shroud of clouds, fogs, rains, and, near the poles, a continual fall of snow. This would be balanced by a constant flow of the currents of the ocean from each pole towards the equator. We should have an excessive circulation of moisture, but no sunshine, and probably only minute changes in the intensity and appearances of one eternal drizzle or shower.

It is plain that this state of things would but ill answer the ends of vegetable and animal life: so that even if the lungs of animals and the leaves of plants were so constructed as to breathe steam instead of air, an atmosphere of unmixed

* Daniell. Meteor. Ess. p. 56.

H

steam would deprive those creatures of most of the
other external conditions of their well-being.

The real state of things which we enjoy, the
steam being mixed in our breath and in our sky
in a moderate quantity, gives rise to results very
different from those which have been described.
The machinery by which these results are produced is not a little curious. It is in fact the
machinery of the *weather*, and therefore the
reader will not be surprised to find it both complex and apparently uncertain in its working.
At the same time some of the general principles
which govern it seem now to be pretty well made
out, and they offer no small evidence of beneficent arrangement.

Besides our atmosphere of aqueous vapour, we
have another and far larger atmosphere of common air; a *permanently elastic* fluid, that is, one
which is not condensed into a liquid form by pressure or cold, such as it is exposed to in the order
of natural events. The pressure of the dry air is
about 29½ inches of mercury; that of the watery
vapour, perhaps, half an inch. Now if we had
the earth quite dry, and covered with an atmosphere of dry air, we can trace in a great
measure what would be the results, supposing
still the equatorial zone to be hot, and the temperature of the surface to decrease perpetually
as we advance into higher latitudes. The air at
the equator would be rarefied by the heat, and

would be perpetually displaced below by the denser portions which belonged to cooler latitudes. We should have a current of air from the equator to the poles in the higher regions of the atmosphere, and at the surface a returning current setting towards the equator to fill up the void so created. Such aerial currents, combined with the rotatory motion of the earth, would produce oblique winds; and we have in fact instances of winds so produced, in the trade winds, which between the tropics blow constantly from the quarters between east and north, and are, we know, balanced by opposite currents in higher regions. The effect of a heated surface of land would be the same as that of the heated zone of the equator, and would attract to it a sea breeze during the day time, a phenomenon, as we also know, of perpetual occurrence.

Now a mass of dry air of such a character as this, is by far the dominant part of our atmosphere; and hence carries with it in its motions the thinner and smaller eddies of aqueous vapour. The latter fluid may be considered as permeating and moving in the interstices of the former, as a spring of water flows through a sand rock.* The lower current of air is, as has been said, directed towards the equator, and hence it resists the motion of the steam, the tendency of which is

* Daniell. p. 129.

in the opposite direction; and prevents or much retards that continual flow of hot vapour into colder regions, by which a constant precipitation would take place in the latter situations.

If, in this state of things, the flow of the current of air, which blows from any colder place into a warmer region, be retarded or stopped, the aqueous vapours will now be able to make their way to the colder point, where they will be precipitated in clouds or showers.

Thus, in the lower part of the atmosphere, there are tendencies to a current of air in one direction, and a current of vapour in the opposite; and these tendencies exist in the average weather of places situated at a moderate distance from the equator. The air tends from the colder to the warmer parts, the vapour from the warmer to the colder.

The various distribution of land and sea, and many other causes, make these currents far from simple. But in general the air current pre-dominates, and keeps the skies clear and the moisture dissolved. Occasional and irregular occurrences disturb this predominance; the mois-ture is then precipitated, the skies are clouded, and the clouds may descend in copious rains.

These alternations of fair weather and showers appear to be much more favourable to vegetable and animal life than any uniform course of weather could have been. To produce this

variety, we have two antagonist forces, by the
struggle of which such changes occur. Steam
and air, two transparent and elastic fluids, ex-
pansible by heat, are in many respects and properties very like each other. Yet, the same heat
similarly applied to the globe, produces at the
surface currents of these fluids, tending in opposite directions. And these currents mix and
balance, conspire and interfere, so that our trees
and fields have alternately water and sunshine ;
our fruits and grain are successively developed
and matured. Why should such laws of heat and
elastic fluids so obtain, and be so combined ? Is
it not in order that they may be fit for such offices ?
There is here an arrangement, which no chance
could have produced. The details of this apparatus may be beyond our power of tracing ; its
springs may be out of our sight. Such circumstances do not make it the less a curious and
beautiful contrivance : they need not prevent our
recognizing the skill and benevolence which we
can discover.

2. But we have not yet done with the machinery of the weather. In ascending from the
earth's surface through the atmosphere, we find
a remarkable difference in the heat and in the
pressure of the air. It becomes much colder, and
much lighter ; men's feelings tell them this ; and
the thermometer and barometer confirm these in-

dications.   And here again we find something to
remark.

In both the simple atmospheres of which we
have spoken, the one of air and the one of steam,
the property which we have mentioned must
exist.   In each of them, both the temperature
and the tension would diminish in ascending.   But
they would diminish at very different rates.   The
temperature, for instance, would decrease much
more rapidly for the same height in dry air than
in steam.   If we begin with a temperature of 80
degrees at the surface, on ascending 5,000 feet
the steam is still 76½ degrees, the air is only 64½
degrees ; at 10,000 feet, the steam is 73 degrees,
the air 48½ degrees ; at 15,000 feet, steam is at
70 degrees, air has fallen below the freezing point
to 31½ degrees.   Hence these two atmospheres
cannot exist together without modifying one another : one must heat or cool the other, so that
the coincident parts may be of the same temperature.   This accordingly does take place, and this
effect influences very greatly the constitution of
the atmosphere.   For the most part, the steam is
compelled to accommodate itself to the temperature of the air, the latter being of much the greater
bulk.   But if the upper parts of the aqueous
vapour be cooled down to the temperature of the
air, they will not by any means exert on the lower
parts of the same vapour so great a pressure as

the gaseous form of these could bear. Hence, there will be a deficiency of moisture in the lower part of the atmosphere, and if water exist there, it will rise by evaporation, the surface feeling an insufficient tension ; and there will thus be a fresh supply of vapour upwards. As, however, the upper regions. already contain as much as their temperature will support in the state of gas, a precipitation will now take place, and the fluid thus formed will descend till it arrives in a lower region, where the tension and temperature are again adapted to its evaporation.

Thus, we can have no equilibrium in such an atmosphere, but a perpetual circulation of vapour between its upper and lower parts. The currents of air which move about in different directions, at different altitudes, will be differently charged with moisture, and as they touch and mingle, lines of cloud are formed, which grow and join, and are spread out in floors, or rolled together in piles. These, again, by an additional accession of humidity, are formed into drops, and descend in showers into the lower regions, and if not evaporated in their fall, reach the surface of the earth.

The varying occurrences thus produced, tend to multiply and extend their own variety. The ascending streams of vapour carry with them that *latent heat* belonging to their gaseous state, which, when they are condensed, they give out as sensible heat. They thus raise the temperature of the upper regions of air, and occasion changes in the pressure and motion of its currents. The clouds, again, by shading the surface of the earth from the sun, diminish the evaporation by which their own substance is supplied, and the heating effects by which currents are caused. Even the mere mechanical effects of the currents of fluid on the distribution of its own pressure, and the dynamical conditions of its motion, are in a high degree abstruse in their principles and complex in their results. It need not be wondered, therefore, if the study of this subject is very difficult and entangled, and our knowledge, after all, very imperfect.

In the middle of all this apparent confusion, however, we can see much that we can understand. And, among other things, we may notice some of the consequences of the difference of the laws of temperature followed by steam and by air in going upwards. One important result is that the atmosphere is much drier, near the surface, than it would have been if the laws of density and temperature had been the same for both gases. If this had been so, the air would always have been *saturated* with vapour. It would have contained as much as the existing temperature could support, and the slightest cooling of any object would have covered it with a watery film like dew. As it is, the air contains much less than its full quantity of vapour : we may often cool an object 10, 20, or 30 degrees without obtaining a deposition of water upon it, or reaching the *dew-point*, as it is called. To have had such a *dripping* state of the atmosphere as the former arrangement would have produced, would have been inconvenient, and so far as we can judge, unsuited to vegetables as well as animals. No evaporation from the surface of either could have taken place under such conditions.

The sizes and forms of clouds appear to depend on the same circumstance, of the air not being saturated with moisture. And it is seemingly much better that clouds should be comparatively small and well defined, as they are, than that they should fill vast depths of the atmosphere with a thin mist, which would have been the consequence of the imaginary condition of things just mentioned.

Here then we have another remarkable exhibition of two laws, in two nearly similar gaseous fluids, producing effects alike in kind, but different in degree, and by the *play* of their difference giving rise to a new set of results, peculiar in their nature and beneficial in their tendency. The *form* of the laws of air and of steam with regard to heat might, so far as we can see, have been more similar, or more dissimilar, than it now is : the rate of each law might have had a different amount from its present one, so as quite to alter the relation of the two. By the laws having such forms and such rates as they have, effects are produced, some of which we can distinctly perceive to be beneficial. Perhaps most persons will feel a strong persuasion, that if we understood the operation of these laws more distinctly, we should see still more clearly the beneficial tendency of these effects, and should probably discover others, at present concealed in the apparent perplexity of the subject.

3. From what has been said, we may see, in a general way, both the causes and the effects of *winds*. They arise from any disturbance by temperature, motion, pressure, &c. of the equilibrium of the atmosphere, and are the efforts of nature to restore the balance. Their office in the economy of nature is to carry heat and moisture from one tract to another, and they are the great agents in the distribution of temperature and the changes of weather. Other purposes might easily be ascribed to them in the business of the vegetable and animal kingdoms, and in the arts of human life, of which we shall not here treat. That character in which we now consider them, that of the machinery of atmospheric changes, and thus, immediately or remotely, the instruments of atmospheric influences, cannot well be refused them by any person.

4. There is still one reflexion which ought not to be omitted. All the changes of the weather,

even the most violent tempests and torrents of rain, may be considered as oscillations about the mean or average condition belonging to each place.   All these oscillations are limited and transient ; the storm spends its fury, the inundation passes off, the sky clears, the calmer course of nature succeeds.   In the forces which produce this derangement, there is a provision for making it short and moderate.   The oscillation stops of itself, like the rolling of a ship, when no longer impelled by the wind.   Now, why should this be so ?   Why should the oscillations, produced by the conflict of so many laws, seemingly quite unconnected with each other, be of this converging and subsiding character ?   Would it be so under all arrangements ?   Is it a matter of mechanical necessity that disturbance must end in the restoration of the medium condition?   By no means. There may be an utter subversion of the equilibrium.   The ship may roll too far, and may *capsize*.   The oscillations may go on, becoming larger and larger, till all trace of the original condition is lost ; till new forces of inequality and disturbance are brought into play ; and disorder and irregularity may succeed, without apparent limit or check in its own nature, like the spread of a conflagration in a city.   This is a possibility in any combination of mechanical forces ; why does it not happen in the one now before us ?   By what good fortune are the powers of heat, of

water, of steam, of air, the effects of the earth s
annual and diurnal motions, and probably other
causes, so adjusted, that through all their struggles
the elemental world goes on, upon the whole, so
quietly and steadily ?   Why is the whole fabric
of the weather never utterly deranged, its balance
lost irrecoverably ?   Why is there not an eternal
conflict, such as the poets imagine to take place
in their chaos ?

> " For Hot, Cold, Moist, and Dry, four champions fierce,
> Strive here for mastery, and to battle bring
> Their embryon atoms :—
>   to whom these most adhere,
> He rules *a moment :* Chaos umpire sits,
> And by decision more embroils the fray."*

A state of things something like that which
Milton here seems to have imagined, is, so far as
we know, not mechanically impossible.   It might
have continued to obtain, if Hot and Cold, and
Moist and Dry had not been compelled to " run
into their places." It will be hereafter seen,
that in the comparatively simple problem of
the solar system, a number of very peculiar
adjustments were requisite, in order that the
system might retain a permanent form, in order
that its motions might have their cycles, its
perturbations their limits and period.   The problem of the continuation of such laws and materials

* Par. Lost, b. II.

as enter into the constitution of the atmosphere, is one manifestly of much greater complexity, and indeed to us probably of insurmountable difficulty as a mechanical problem. But all that investigation and analogy teach us, tends to shew that it will resemble the other problem in the nature of its result; and that certain relations of its data, and of the laws of its elements, are necessary requisites, for securing the stability of its mean condition, and for giving a small and periodical character to its deviations from such a condition.

It would then be probable, from this reflexion alone, that in determining the quantity and the law and intensity of the forces, of earth, water, air, and heat, the same regard has been shewn to the permanency and stability of the terrestrial system, which may be traced in the adjustment of the masses, distances, positions, and motions of the bodies of the celestial machine.

This permanency appears to be, of itself, a suitable object of contrivance. The purpose for which the world was made could be answered only by its being preserved. But it has appeared, from the preceding part of this and the former chapter, that this permanence is a permanence of a state of things adapted by the most remarkable and multiplied combinations to the well-being of man, of animals, of vegetables. The adjustments and conditions therefore, beyond

the reach of our investigation as they are, by
which its permanence is secured, must be conceived as fitted to add, in each of the instances
above adduced, to the admiration which the
several manifestations of Intelligent Beneficence
are calculated to excite.

CHAPTER XI.

The Laws of Electricity.

ELECTRICITY undoubtedly exists in the atmosphere in most states of the air ; but we know very
imperfectly the laws of this agent, and are still
more ignorant of its atmospheric operation.   The
present state of science does not therefore enable
us to perceive those adaptations of its laws to its
uses, which we can discover in those cases where
the laws and the uses are both of them more
apparent.

We can, however, easily make out that electrical agency plays a very considerable part
among the clouds, in their usual conditions and
changes.   This may be easily shewn by Frank-
lin s experiment of the electrical kite. The clouds
are sometimes positively, sometimes negatively,
charged, and the rain which descends from them
offers also indications of one or other kind of

electricity. The changes of wind and alterations of the form of the clouds are generally accompanied with changes in these electrical indications. Every one knows that a thunder-cloud is strongly charged with the electric fluid, (if it be a fluid,) and that the stroke of the lightning is an electrical discharge. We may add that it appears, by recent experiments, that a transfer of electricity between plants and the atmosphere is perpetually going on during the process of vegetation.

We cannot trace very exactly the precise circumstances, in the occurrences of the atmospheric regions, which depend on the influence of the laws of electricity : but we are tolerably certain, from what has been already noticed, that if these laws did not exist, or were very different from what they now are, the action of the clouds and winds, and the course of vegetation, would also be other than it now is.

It is therefore at any rate very probable that electricity has its appointed and important purposes in the economy of the atmosphere. And this being so, we may see a use in the thunder-storm and the stroke of the lightning. These violent events are, with regard to the electricity of the atmosphere, what winds are with regard to heat and moisture. They restore the equilibrium where it has been dissolved, and carry the fluid from places where it is superfluous, to others where it is deficient.

We are so constituted, however, that these crises impress almost every one with a feeling of awe. The deep lowering gloom of the thunder-cloud, the overwhelming burst of the explosion, the flash from which the steadiest eye shrinks, and the irresistible arrow of the lightning which no earthly substance can withstand, speak of something fearful, even independently of the personal danger which they may whisper. They convey, far more than any other appearance does, the idea of a superior and mighty power, manifesting displeasure and threatening punish-ment. Yet we find that this is not the language which they speak to the physical enquirer : he sees these formidable symptoms only as the means or the consequences of good. What office the thunderbolt and the whirlwind may have in the *moral* world, we cannot here discuss : but certainly *he* must speculate as far beyond the limits of philosophy as of piety, who pretends to have learnt that there their work has more of evil than of good. In the *natural* world, these ap-parently destructive agents are, like all the other movements and appearances of the atmosphere, parts of a great scheme, of which every discover-able purpose is marked with beneficence as well as wisdom.

CHAPTER XII.

The Laws of Magnetism.

MAGNETISM has no very obvious or apparently extensive office in the mechanism of the atmosphere and the earth : but the mention of it may be introduced, because its ascertained relations to the other powers which exist in the system are well suited to shew us the connexion subsisting throughout the universe, and to check the suspicion, if any such should arise, that any law of nature is without its use. The parts of creation when these uses are most obscure, are precisely those parts when the laws themselves are least known.

When indeed we consider the vast service of which magnetism is to man, by supplying him with that invaluable instrument the mariner's compass, many persons will require no further evidence of this property being introduced into the frame of things with a worthy purpose. As however, we have hitherto excluded *use in the arts* from our line of argument, we shall not here make an exception in favour of navigation, and what we shall observe belongs to another view of the subject.

Magnetism has been discovered in modern

times to have so close a connexion with galvanism, that they may be said to be almost different aspects of the same agent. All the phenomena which we can produce with magnets, we can imitate with coils of galvanic wire. That galvanism exists in the earth, we need no proof. Electricity, which appears to be only galvanism in equilibrium, is there in abundance; and recently, Mr. Fox* has shewn by experiment that metalliferous veins, as they lie in the earth, exercise a galvanic influence on each other. Something of this kind might have been anticipated; for masses of metal in contact, if they differ in temperature or other circumstances, are known to produce a galvanic current. Hence we have undoubtedly streams of galvanic influence moving along in the earth. Whether or not such causes as these produce the directive power of the magnetic needle, we cannot here pretend to decide; they can hardly fail to affect it. The Aurora Borealis too, probably an electrical phenomenon, is said, under particular circumstances, to agitate the magnetic needle. It is not surprising, therefore, that, if electricity have an important office in the atmosphere, magnetism should exist in the earth. It seems likely, that the magnetic properties of the earth may be collateral results of the existence of the same

* Phil. Trans. 1831.

cause by which electrical agency operates; an agency which, as we have already seen, has important offices in the processes of vegetable life. And thus magnetism belongs to the same system of beneficial contrivance to which electricity has been already traced.

We see, however, on this subject very dimly and a very small way. It can hardly be doubted that magnetism has other functions than those we have noticed.

CHAPTER XIII.

The Properties of Light with regard to Vegetation.

THE illuminating power of light will come under our consideration hereafter. Its agency, with regard to organic life, is too important not to be noticed, though this must be done briefly. Light appears to be as necessary to the health of plants as air or moisture. A plant may, indeed, grow without it, but it does not appear that a species could be so continued. Under such a privation, the parts which are usually green, assume a white colour, as is the case with vegetables grown in a cellar, or protected by a covering for the sake of producing this very effect; thus, celery is in this manner blanched, or *etiolated.*

The part of the process of vegetable life for which light is especially essential, appears to be the functions of the leaves ; these are affected by this agent in a very remarkable manner. The moisture which plants imbibe is, by their vital energies, carried to their leaves ; and is then brought in contact with the atmosphere, which, besides other ingredients, contains, in general, a portion of carbonic acid. *So long as light is present*, the leaf decomposes the carbonic acid, appropriates the carbon to the formation of its own proper juices, and returns the disengaged oxygen into the atmosphere ; thus restoring the atmospheric air to a condition in which it is more fitted than it was before for the support of animal life. The plant thus prepares the support of life for other creatures at the same time that it absorbs its own. The greenness of those members which affect that colour, and the disengagement of oxygen, are the indications that its vital powers are in healthful action : as soon as we remove light from the plant, these indications cease : it has no longer power to imbibe carbon and disengage oxygen, but on the contrary, it gives back some of the carbon already obtained, and robs the atmosphere of oxygen for the purpose of reconverting this into carbonic acid.

It cannot well be conceived that such effects of light on vegetables, as we have described, should occur, if that agent, of whatever nature it is, and those organs, had not been adapted to each other. But the subject is here introduced that the reader may the more readily receive the conviction of combining purpose which must arise, on finding that an agent, possessing these very peculiar chemical properties, is employed to produce also those effects of illumination, vision, &c., which form the most obvious portion of the properties of light.

Chapter XIV.

Sound.

Besides the function which air discharges as the great agent in the changes of meteorology and vegetation, it has another office, also of great and extensive importance, as the vehicle of sound.

1. The communication of sound through the air takes place by means of a process altogether different from anything of which we have yet spoken : namely, by the propagation of minute *vibrations* of the particles from one part of the fluid mass to another, without any local motion of the fluid itself.

Perhaps we may most distinctly conceive the kind of effect here spoken of, by comparing it to the motion produced by the wind in a field of standing corn ; grassy waves travel visibly over the field, in the direction in which the wind blows, but this appearance of an object moving is delusive. The only real motion is that of the ears of grain, of which each goes and returns, as the stalk stoops and recovers itself. This motion affects *successively* a line of ears in the direction of the wind, and affects *simultaneously* all those ears of which the elevation or depression forms one visible wave. The elevations and depressions are propagated in a constant direction, while the parts with which the space is filled only vibrate to and fro. Of exactly such a nature is the propagation of sound through the air. The particles of air go and return through very minute spaces, and this vibratory motion runs through the atmosphere from the sounding body to the ear. Waves, not of elevation and depression, but of condensation and rarefaction, are transmitted ; and the sound thus becomes an object of sense to the organ.

Another familiar instance of the propagation of vibrations we have in the circles on the surface of smooth water, which diverge from the point where it is touched by a small object, as a drop of rain. In the beginning of a shower for instance, when the drops come distinct, though frequent, we may see each drop giving rise to a ring, formed of two or three close concentric circles, which grow and spread, leaving the interior of the circles smooth, and gradually reaching parts of the surface more and more distant from their origin. In this instance, it is clearly not a portion of the water which flows onwards ; but the disturbance, the rise and fall of the surface which makes the ring-formed waves, passes into wider and wider circles, and thus the undulation is transmitted from its starting-place, to points in all directions on the surface of the fluid.

The diffusion of these ring-formed undulations from their centre resembles the diffusion of a sound from the place where it is produced to the points where it is heard. The disturbance, or vibration, by which it is conveyed, travels at the same rate in all directions, and the waves which are propagated are hence of a circular form. They differ, however, from those on the surface of water ; for sound is communicated upwards and downwards, and in all intermediate directions, as well as horizontally ; hence the waves of sound are spherical, the point where the sound is produced being the centre of the sphere.

This diffusion of vibrations in spherical shells of successive condensation and rarefaction, will easily be seen to be different from any local motion of the air, as wind, and to be independent of that. The circles on the surface of water will spread on a river which is flowing, provided it be smooth, as well as on a standing canal.

Not only are such undulations propagated almost undisturbed by any local motion of the fluid in which they take place, but also, many may be propagated in the same fluid at the same time, without disturbing each other. We may see this effect on water. When several drops fall near each other, the circles which they produce cross each other, without either of them being lost, and the separate courses of the rings may still be traced.

All these consequences, both in water, in air, and in any other fluid, can be very exactly investigated upon mechanical principles, and the greater part of the phenomena can thus be shewn to result from the properties of the fluids.

There are several remarkable circumstances in the way in which air answers its purpose as the vehicle of sound, of which we will now point out a few.

2. The *loudness* of sound is such as is convenient for common purposes. The organs of speech can, in the present constitution of the air, produce, without fatigue, such a tone of voice as can be heard with distinctness and with comfort. That any great alteration in this element might be incommodious, we may judge from the difficulties to which persons are subject who are dull of hearing, and from the disagreeable effects of a voice much louder than usual, or so low as to be indistinct. Sounds produced by the human organs, with other kinds of air, are very different from those in our common air If a man inhale

a quantity of hydrogen gas, and then speak, his voice is scarcely audible.

The loudness of sounds become smaller in proportion as they come from a greater distance. This enables us to judge of the distance of objects, in some degree at least, by the sounds which proceed from them. Moreover it is found that we can judge of the position of objects by the ear : and this judgment seems to be formed by comparing the loudness of the impression of the same sound on the two ears and two sides of the head.*

The loudness of sounds appears to depend on the *extent* of vibration of the particles of air, and this is determined by the vibrations of the sounding body.

3. The *pitch*, or the *differences of acute and grave*, in sounds, form another important property, and one which fits them for a great part of their purposes. By the succession of different *notes*, we have all the results of melody and harmony in musical sound ; and of intonation and mdoulation of the voice, of accent, cadence, emphasis, expression, passion, in speech. The song of birds, which is one of their principal modes of communication, depends chiefly for its distinctions and its significance upon the combinations of acute and grave.

* Mr. Gough in Manch. Mem. vol. v.

These differences are produced by the different *rapidity* of vibration of the particles of air. The gravest sound has about eighty vibrations in a second, the most acute about one thousand. Between these limits each sound has a musical character, and from the different relations of the number of vibrations in a second arise all the differences of musical intervals, concords and discords.

4. The *quality* of sounds is another of their differences. This is the name given to the difference of notes of the same pitch, that is the same note as to acute and grave, when produced by different instruments. If a flute and a violin be in unison, the notes are still quite different sounds. It is this kind of difference which distinguishes the voice of one man from that of another: and it is manifestly therefore one of great consequence; since it connects the voice with the particular person, and is almost necessary in order that language may be a medium of intercourse between men.

5. The *articulate* character of sounds is for us one of the most important arrangements which exist in the world; for it is by this that they become the interpreters of thought, will and feeling, the means by which a person can convey his wants, his instructions, his promises, his kindness, to others; by which one man can regulate the actions and influence the convictions and judgments of another. It is in virtue of the pos-

sibility of shaping air into words, that the imperceptible vibrations which a man produces in the atmosphere, become some of his most important actions; the foundations of the highest moral and social relations ; and the condition and instrument of all the advancement and improvement of which he is susceptible.

It appears that the differences of articulate sound arise from the different form of the cavity through which the sound is made to proceed immediately *after* being produced. In the human voice the sound is produced in the larynx, and modified by the cavity of the mouth, and the various organs which surround this cavity. The laws by which articulate sounds are thus produced have not yet been fully developed, but appear to be in the progress of being so.

The properties of sounds which have been mentioned, differences of loudness, of pitch, of quality, and articulation, appear to be all requisite in order that sound shall answer its purposes in the economy of animal and of human life. And how was the air made capable of conveying these four differences, at the same time that the organs were made capable of producing them ? Surely by a most refined and skilful adaptation, applied with a most comprehensive design.

6. Again; is it by chance that the air and the *ear* exist together ? Did the air produce the organization of the ear? or the ear, independently organised,

anticipate the constitution of the atmosphere?
Or is not the only intelligible account of the
matter, this, that one was made for the other:
that there is a mutual adaptation produced by an
Intelligence which was acquainted with the properties of both; which adjusted them to each other
as we find them adjusted, in order that birds might
communicate by song, that men might speak and
hear, and that language might play its extraordinary part in its operation upon men's thoughts,
actions, institutions, and fortunes?

The vibrations of an elastic fluid like the air,
and their properties, follow from the laws of motion;
and whether or not these laws of the motion of fluids
might in reality have been other than they are,
they appear to us inseparably connected with the
existence of matter, and as much a thing of
necessity as we can conceive any thing in the
universe to be. The propagation of such vibrations, therefore, and their properties, we may at
present allow to be a necessary part of the
constitution of the atmosphere. But what is it
that makes these vibrations become sound? How
is it that they produce such an effect on our
senses, and, through those, on our minds? The
vibrations of the air seem to be of themselves no
more fitted to produce sound, than to produce
smell. We know that such vibrations do not
universally produce sound, but only between
certain limits. When the vibrations are fewer

than eighty in a second, they are perceived as separate throbs, and not as a continued sound; and there is a certain limit of rapidity, beyond which the vibrations become inaudible. This limit is different to different ears, and we are thus assured by one person's ear that there are vibrations, though to that of another they do not produce sound. How was the human ear adapted so that its perception of vibrations as sounds should fall within these limits?—the very limits within which the vibrations fall, which it most concerns us to perceive: those of the human voice for instance? How nicely are the organs adjusted with regard to the most minute mechanical motions of the elements!

CHAPTER XV.

The Atmosphere.

WE have considered in succession a number of the properties and operations of the atmosphere, and have found them separately very curious. But an additional interest belongs to the subject when we consider them as combined. The atmosphere under this point of view must appear a contrivance of the most extraordinary kind. To answer any of its purposes, to carry on any of its processes, separately, requires peculiar

arrangements and adjustments ; to answer, all at
once, purposes so varied, to combine without
confusion so many different trains, implies powers
and attributes which can hardly fail to excite in
a high degree our admiration and reverence.

If the atmosphere be considered as a vast
machine, it is difficult to form any just conception of the profound skill and comprehensiveness
of design which it displays.   It diffuses and
tempers the heat of different climates ; for this
purpose it performs a circulation occupying the
whole range from the pole to the equator ; and
while it is doing this, it executes many smaller
circuits between the sea and the land.   At the
same time, it is the means of forming clouds and
rain, and for this purpose, a perpetual circulation
of the watery part of the atmosphere goes on between its lower and upper regions.   Besides this
complication of circuits, it exercises a more irregular agency, in the occasional winds which blow
from all quarters, tending perpetually to restore
the equilibrium of heat and moisture.   But this
incessant and multiplied activity discharges only
a part of the functions of the air.   It is, moreover,
the most important and universal material of the
growth and sustenance of plants and animals; and
is for this purpose every where present and almost
uniform in its quantity.   With all its local motion,
it has also the office of a medium of communication between intelligent creatures, which office it

performs by another set of motions, entirely different both from the circulation and the occasional movements already mentioned; these different kinds of motions not interfering materially with each other : and this last purpose, so remote from the others in its nature, it answers in a manner so perfect and so easy, that we cannot imagine that the object could have been more completely attained, if this had been the sole purpose for which the atmosphere had been created. With all these qualities, this extraordinary part of our terrestrial system is scarcely ever in the way : and when we have occasion to do so, we put forth our hand and push it aside, without being aware of its being near us.

We may add, that it is, in addition to all that we have hitherto noticed, a constant source of utility and beauty in its effects on light. Without air we should see nothing, except objects on which the sun's rays fell, directly or by reflection. It is the atmosphere which converts sunbeams into daylight, and fills the space in which we are with illumination.

The contemplation of the atmosphere, as a machine which answers all these purposes, is well suited to impress upon us the strongest conviction of the most refined, far-seeing, and far-ruling contrivance. It seems impossible to suppose that these various properties were so bestowed and so combined, any otherwise than by a beneficent and

intelligent Being, able and willing to diffuse
organization, life, health, and enjoyment through
all parts of the visible world ; possessing a fertility
of means which no multiplicity of objects could
exhaust, and a discrimination of consequences
which no complication of conditions could embarrass.

Chapter XVI.

Light.

BESIDES the hearing and sound there is another
mode by which we become sensible of the impressions of external objects, namely, sight and
light.   This subject also offers some observations
bearing on our present purpose.

It has been declared by writers on Natural
Theology, that the human eye exhibits such
evidence of design and skill in its construction,
that no one, who considers it attentively, can resist
this impression :   nor does this appear to be
saying too much.   It must, at the same time, be
obvious that this construction of the eye could
not answer its purposes, except the constitution
of light corresponded to it.   Light is an element
of the most peculiar kind and properties, and such
an element can hardly be conceived to have been
placed in the universe without a regard to its
operation and functions.   As the eye is made for

light, so light must have been made, at least among other ends, for the eye.

1. We must expect to comprehend imperfectly only the mechanism of the elements. Still, we have endeavoured to shew that in some instances the arrangements by which their purposes are effected are, to a certain extent, intelligible. In order to explain, however, in what manner light answers those ends which appear to us its principal ones, we must know something of the nature of light. There have, hitherto, been, among men of science, two prevailing opinions upon this subject : some considering light as consisting in the emission of luminous particles ; others accounting for its phenomena by the propagation of vibrations through a highly subtle and elastic *ether*. The former opinion has, till lately, been most generally entertained in this country, having been the hypothesis on which Newton made his calculations ; the latter is the one to which most of those persons have been led, who, in recent times, have endeavoured to deduce general conclusions from the newly discovered phenomena of light. Among these persons, the *theory of undulations* is conceived to be established in nearly the same manner, and almost as certainly, as the doctrine of universal gravitation ; namely, by a series of laws inferred from numerous facts, which, proceeding from different sets of phenomena, are found to converge to one common view ; and by

calculations founded upon the theory, which, indicating new and untried facts, are found to agree exactly with experiment.

We cannot here introduce a sketch of the progress by which the phenomena have thus led to the acceptance of the theory of undulations. But this theory appears to have such claims to our assent, that the views which we have to offer with regard to the design exercised in the adaptation of light to its purposes, will depend on the undulatory theory, so far as they depend on theory at all.*

2. The impressions of sight, like those of hearing, differ in intensity and in kind. *Brightness* and *Colour* are the principal differences among visible things, as loudness and pitch are among sounds. But there is a singular distinction between these senses in one respect : every object and part of an object seen, is necessarily and inevitably referred to some *position* in the space before us ; and hence visible things have place, magnitude, form, as well as light, shade, and colour. There is nothing analogous to this in the sense of hearing ; for though we can, in some approximate degree, *guess* the situation of the point from which a sound proceeds, this is a

* The reader who is acquainted with the two theories of light, will perceive that though we have adopted the doctrine of the ether, the greater part of the arguments adduced would be equally forcible, if expressed in the language of the theory of emission.

secondary process, distinguishable from the perception of the sound itself; whereas we cannot conceive visible things without form and place.

The law according to which the sense of vision is thus affected, appears to be this. By the properties of light, the external scene produces, through the transparent parts of the eye, an image or picture exactly resembling the reality, upon the back part of the retina : and each point which we see is seen in the direction of a line passing from its image on the retina, through the centre of the pupil of the eye.* In this manner we perceive by the eye the situation of every point, at the same time that we perceive its existence ; and by combining the situations of many points, we have forms and outlines of every sort.

That we should receive from the eye this notice of the position of the object as well as of its other visible qualities, appears to be absolutely necessary for our intercourse with the external world ; and the faculty of doing so is so intimate a part of our constitution that we cannot conceive ourselves divested of it. Yet in order to imagine ourselves destitute of this faculty, we have only to suppose that the eye should receive its impressions as the ear does, and should apprehend

* Or rather through the *focal centre* of the eye, which is always near the centre of the pupil.

red and green, bright and dark, without placing
them side by side; as the ear takes in the different sounds which compose a concert, without
attributing them to different parts of space.

The peculiar property thus belonging to vision,
of perceiving position, is so essential to us, that
we may readily believe that some particular
provision has been made for its existence. The
remarkable mechanism of the eye (precisely resembling that of a *camera obscura*,) by which it
produces an image on the nervous web forming
its hinder part, seems to have this effect for its
main object. And this mechanism necessarily
supposes certain corresponding properties in light
itself, by means of which such an effect becomes
possible.

The main properties of light which are concerned in this arrangement, are *reflexion* and
refraction: reflexion, by which light is reflected
and scattered by all objects, and thus comes to
the eye from all: and refraction, by which its
course is bent, when it passes obliquely out of
one transparent medium into another; and by
which, consequently, convex transparent substances, such as the cornea and humours of the
eye, possess the power of making the light converge to a *focus* or point; an assemblage of such
points forming the images on the retina, which
we have mentioned.

Reflexion and refraction are therefore the essential and indispensable properties of light ; and so far as we can understand, it appears that it was necessary that light should · possess such properties, in order that it might form a medium of communication between man and the external world. We may consider its power of passing through transparent media (as air) to be given in order that it may enlighten the earth ; its affection of reflexion, for the purpose of making colours visible ; and its refraction to be bestowed, that it may enable us to discriminate figure and position, by means of the lenses of the eye.

In this manner light may be considered as constituted with a peculiar reference to the eyes of animals, and its leading properties may be looked upon as contrivances or adaptations to fit it for its visual office. And in such a point of view the perfection of the contrivance or adaptation must be allowed to be very remarkable.

3. But besides the properties of reflexion and refraction, the most obvious laws of light, an extraordinary variety of phenomena have lately been discovered, regulated by other laws of the most curious kind, uniting great complexity with great symmetry. We refer to the phenomena of diffraction, polarisation, and periodical colours, produced by crystals and by thin plates. We have, in these facts, a vast mass of properties and laws, offering a subject of study which has been pursued with eminent skill and intelligence. But

these properties and laws, so far as has yet been discovered, exert no agency whatever, and have no purpose, in the general economy of nature. Beams of light polarised in contrary directions exhibit the most remarkable differences when they pass through certain crystals, but manifest no discoverable difference in their immediate impression on the eye. We have, therefore, here, a number of laws of light, which we cannot perceive to be established with any design which has a reference to the other parts of the universe.

Undoubtedly it is exceedingly possible that these differences of light may operate in some quarter, and in some way, which we cannot detect; and that these laws may have purposes and may answer ends of which we have no suspicion. All the analogy of nature teaches us a lesson of humility, with regard to the reliance we are to place on our discernment and judgment as to such matters. But with our present knowledge, we may observe, that this curious system of phenomena appears to be a collateral result of the mechanism by which the effects of light are produced; and therefore a necessary consequence of the existence of that element of which the offices are so numerous and so beneficent.

The new properties of light, and the speculations founded upon them, have led many persons to the belief of the undulatory theory; which, as we have said, is considered by some philosophers as demonstrated. If we adopt this theory, we consider the luminiferous ether to have no local motion; and to produce refraction and reflexion by the operation of its elasticity alone. We must necessarily suppose the tenuity of the ether to be extreme; and if we moreover suppose its tension to be very great, which the vast velocity of light requires us to suppose, the vibrations by which light is propagated will be *transverse* vibrations, that is the motion to and fro will be athwart the line along which the undulation travels; and from this circumstance all the laws of polarisation necessarily follow. And the properties of transverse vibrations, combined with the properties of vibrations in general, give rise to all the curious and numerous phenomena of colours of which we have spoken.

If the vibrations be transverse, they may be resolved into two different planes; this is *polarisation:* if they fall on a medium which has different elasticity in different directions, they will be divided into two sets of vibrations; this is *double refraction;* and so on. Some of the new properties, however, as the fringes of shadows and the colours of thin plates, follow from the undulatory theory, whether the vibrations be transverse or not.

It would appear, therefore, that the propagation of light by means of a subtle medium, leads necessarily to the extraordinary collection of

properties which have recently been discovered;
and, at any rate, its propagation by the transverse
vibrations of such a medium does lead inevitably
to these results.

Leaving it therefore to future times to point
out the other reasons (or *uses* if they exist) of
these newly discovered properties of light, in
their bearing on other parts of the world, we may
venture to say, that if light was to be propagated
through transparent media by the undulations of a
subtle fluid, these properties must result, as necessarily as the rainbow results from the unequal
refrangibility of different colours. This phenomenon and those, appear alike to be the collateral
consequences of the laws impressed on light with
a view to its principal offices.

Thus the exquisitely beautiful and symmetrical
phenomena and laws of polarisation, and of crystalline and other effects, may be looked upon as indications of the delicacy and subtlety of the
mechanism by which man, through his visual
organs, is put in communication with the external
world; is made acquainted with the forms and
qualities of objects in the most remote regions of
space; and is enabled, in some measure, to
determine his position and relation in a universe
in which he is but an atom.

4. If we suppose it clearly established that
light is produced by the vibrations of an ether,
we find considerations offer themselves, similar to

those which occurred in the case of sound. The
vibrations of this ether affect our organs with the
sense of light and colour. Why, or how do they
do this? It is only within certain limits that the
effect is produced, and these limits are comparatively narrower here than in the case of sound. The
whole scale of colour, from violet to crimson, lies
between vibrations which are 458 million millions,
and 727 million millions in a second ; a proportion
much smaller than the corresponding ratio for
perceptible sounds. Why should such vibrations
produce perception in the eye, and no others?
There must be here some peculiar adaptation of the
sensitive powers to these wonderfully minute and
condensed mechanical motions. What happens
when the vibrations are slower than the red, or
quicker than the blue? They do not produce
vision : do they produce any effect? Have they
any thing to do with heat or with electricity?
We cannot tell. The ether must be as susceptible
of these vibrations, as of those which produce
vision. But the mechanism of the eye is adjusted
to this latter kind only ; and this precise kind,
(whether alone or mixed with others,) proceeds
from the sun and from other luminaries, and
thus communicates to us the state of the visible
universe. The mere material elements then are
full of properties which we can understand no
otherwise, than as the results of a refined contrivance.

CHAPTER XVII.

The Ether.

IN what has just been said, we have spoken of light, only with respect to its power of illuminating objects, and conveying the impression of them to the eye. It possesses, however, beyond all doubt, many other qualities. Light is intimately connected with heat, as we see in the case of the sun and of flame; yet it is clear that light and heat are not identical. Light is evidently connected too with electricity and galvanism; and perhaps, through these, with magnetism: it is, as has already been mentioned, indispensably necessary to the healthy discharge of the functions of vegetable life; without it plants cannot duly exercise their vital powers: it manifests also chemical action in various ways.

The luminiferous *ether* then, if we so call the medium in which light is propagated, must possess many other properties besides those mechanical ones on which the illuminating power depends. It must not be merely like a fluid poured into the vacant spaces and interstices of the material world, and exercising no action on objects; it must affect the physical, chemical and vital powers of what it touches. It must be a great and active agent in the work of the universe, as well as an active reporter of what is done by other agents. It must possess a number of complex and refined contrivances and adjustments which we cannot analyze, bearing upon plants and chemical compounds, and the imponderable agents ; as well as those laws which we conceive that we have analyzed, by which it is the vehicle of illumination and vision.

We have had occasion to point out how complex is the machinery of the atmosphere, and how varied its objects ; since, besides being the means of communication as the medium of sound, it has known laws which connect it with heat and moisture ; and other laws, in virtue of which it is decomposed by vegetables. It appears, in like manner, that the ether is not only the vehicle of light, but has also laws, at present unknown, which connect it with heat, electricity, and other agencies ; and other laws through which it is necessary to vegetables, enabling them to decompose air. All analogy leads us to suppose that if we knew as much of the constitution of the luminiferous ether as we know of the constitution of the atmosphere, we should find it a machine as complex and artificial, as skilfully and admirably constructed.

We know at present very little indeed of the construction of this machine. Its *existence* is, perhaps, satisfactorily made out ; in order that we may not interrupt the progress of our argument, we shall refer to other works for the reasonings which appear to lead to this conclusion. But whether heat, electricity, galvanism, magnetism, be fluids; or effects or modifications of fluids; and whether such fluids or *ethers* be the same with the luminiferous ether, or with each other; are questions of which all or most appear to be at present undecided, and it would be presumptuous and premature here to take one side or the other.

The mere fact, however, that there is such an ether, and that it has properties related to other agents, in the way we have suggested, is well calculated to extend our views of the structure of the universe, and of the resources, if we may so speak, of the Power by which it is arranged. The solid and fluid matter of the earth is the most obvious to our senses; over this, and in its cavities, is poured an invisible fluid, the air, by which warmth and life are diffused and fostered, and by which men communicate with men: over and through this again, and reaching, so far as we know, to the utmost bounds of the universe, is spread another most subtle and attenuated fluid, which, by the play of another set of agents, aids the energies of nature, and which, filling all parts of space, is a means of communication with other planets and other systems.

There is nothing in all this like any material necessity, compelling the world to be as it is and no otherwise. How should the properties of these three great classes of agents, visible objects, air, and light, so harmonize and assist each other, that order and life should be the result. Without all the three, and all the three constituted in their present manner, and subject to their present laws, living things could not exist. If the earth had no atmosphere, or if the world had no ether, all must be inert and dead. Who constructed these three extraordinarily complex pieces of machinery, the earth with its productions, the atmosphere, and the ether? Who fitted them into each other in many parts, and thus made it possible for them to work together? We conceive there can be but one answer; a most wise and good God.

CHAPTER XVIII.

Recapitulation.

1. IT has been shewn in the preceding chapters that a great number of quantities and laws appear to have been *selected* in the construction of the universe; and that by the adjustment to each other of the magnitudes and laws thus selected, the constitution of the world is what we find it, and is fitted for the support of vegetables and animals, in a manner in which it could not have

been, if the properties and quantities of the elements had been different from what they are. We shall here recapitulate the principal of the laws and magnitudes to which this conclusion has been shewn to apply.

1. The Length of the Year, which depends on the force of the attraction of the sun, and its distance from the earth.

2. The Length of the Day.

3. The Mass of the Earth, which depends on its magnitude and density.

4. The Magnitude of the Ocean.

5. The Magnitude of the Atmosphere.

6. The Law and Rate of the Conducting Power of the Earth.

7. The Law and Rate of the Radiating Power of the Earth.

8. The Law and Rate of the Expansion of Water by Heat.

9. The Law and Rate of the Expansion of Water by Cold, below 40 degrees.

10. The Law and Quantity of the Expansion of water in Freezing.

11. The Quantity of Latent Heat absorbed in Thawing.

12. The Quantity of Latent Heat absorbed in Evaporation.

13. The Law and Rate of Evaporation with regard to Heat.

14. The Law and Rate of the Expansion of Air by Heat.

15. The Quantity of Heat absorbed in the Expansion of Air.

16. The Law and Rate of the Passage of Aqueous Vapour through Air.

17. The Laws of Electricity ; its relations to Air and Moisture.

18. The Fluidity, Density, and Elasticity of the Air, by means of which its vibrations produce Sound.

19. The Fluidity, Density, and Elasticity of the Ether, by means of which its vibrations produce Light.

2. These are the *data*, the *elements*, as astronomers call the quantities which determine a planet's orbit, on which the mere *inorganic* part of the universe is constructed. To these, the constitution of the organic world is adapted in innumerable points, by laws of which we can trace the results, though we cannot analyze their machinery. Thus, the vital functions of vegetables have periods which correspond to the length of the year, and of the day ; their vital powers have forces which correspond to the force of gravity ; the sentient faculties of man are such that the vibrations of air, (within certain limits,) are perceived as sound, those of ether, as light. And while we are enumerating these correspondencies,

we perceive that there are thousands of others, and that we can only select a very small number of those where the relation happens to be most clearly made out or most easily explained.

Now, in the list of the mathematical *elements* of the universe which has just been given, why have we such laws and such quantities as there occur, and no other? For the most part, the data there enumerated are independent of each other, and might be altered separately, so far as the mechanical conditions of the case are concerned. Some of these data probably depend on each other. Thus the latent heat of aqueous vapour is perhaps connected with the difference of the rate of expansion of water and of steam. But all natural philosophers will, probably, agree, that there must be, in this list, a great number of things entirely without any mutual dependence, as the year and the day, the expansion of air and the expansion of steam. There are, therefore, it appears, a number of things which, in the structure of the world, might have been otherwise, and which are what they are in consequence of choice or of chance. We have already seen, in many of the cases separately, how unlike chance every thing looks:—that substances, which might have existed any how, so far as they themselves are concerned, exist exactly in such a manner and measure as they should, to secure the welfare of other things: —that the laws are tempered and fitted together in the only way in which the world could have gone on, according to all that we can conceive of it. This must, therefore, be the work of choice; and if so, it cannot be doubted, of a most wise and benevolent Chooser.

3. The appearance of choice is still further illustrated by the variety as well as the number of the laws selected. The laws are unlike one another. Steam certainly expands at a very different *rate* from air by the application of heat, probably according to a different *law:* water expands in freezing, but mercury contracts : heat travels in a manner quite different through solids and fluids. Every separate substance has its own density, gravity, cohesion, elasticity, its relations to heat, to electricity, to magnetism ; besides all its chemical affinities, which form an endless throng of laws, connecting every one substance in creation with every other, and different for each pair anyhow taken. Nothing can look less like a world formed of atoms operating upon each other according to some universal and inevitable laws, than this does : if such a system of things be conceivable, it cannot be our system. We have, it may be, fifty simple substances in the world ; each of which is invested with properties, both of chemical and mechanical action, altogether different from those of any other substance. Every portion, however minute, of any of these, possesses all the properties of the substance.

Of each of these substances there is a certain unalterable quantity in the universe; when combined, their compounds exhibit new chemical affinities, new mechanical laws. Who gave these different laws to the different substances? who proportioned the quantity of each? But suppose this done. Suppose these substances in existence; in contact; in due proportion to each other. Is *this* a world, or at least our world? No more than the mine and the forest are the ship of war or the factory. These elements, with their constitution perfect, and their proportion suitable, are still a mere chaos. They must be put in their places. They must not be where their own properties would place them. They must be made to assume a particular arrangement, or we can have no regular and permanent course of nature. This arrangement must again have additional peculiarities, or we can have no organic portion of the world. The millions of millions of particles which the world contains, must be finished up in as complete a manner, and fitted into their places with as much nicety, as the most delicate wheel or spring in a piece of human machinery. What are the habits of thought to which it can appear possible that this could take place without design, intention, intelligence, purpose, knowledge?

In what has just been said, we have spoken only of the constitution of the inorganic part of the universe. The mechanism, if we may so call it, of vegetable and animal life, is so far beyond our comprehension, that though some of the same observations might be applied to it, we do not dwell upon the subject. We know that in these processes also, the mechanical and chemical properties of matter are necessary, but we know too that these alone will not account for the phenomena of life. There is something more than these. The lowest stage of vitality and irritability appears to carry us beyond mechanism, beyond affinity. All that has been said with regard to the exactness of the adjustments, the combination of various means, the tendency to continuance, to preservation, is applicable with additional force to the organic creation, so far as we can perceive the means employed. These, however, belong to a different province of the subject, and must be left to other hands.

BOOK II.

COSMICAL ARRANGEMENTS.

WHEN we turn our attention to the larger portions of the universe, the sun, the planets, and the earth as one of them, the moon and other satellites, the fixed stars and other heavenly bodies ;—the views which we obtain concerning their mutual relations, arrangement and movements, are called, as we have already stated, *cosmical* views. These views will, we conceive, afford us indications of the wisdom and care of the Power by which the objects which we thus consider, were created and are preserved : and we shall now proceed to point out some circumstances in which these attributes may be traced.

It has been observed by writers on Natural Theology, that the arguments for the being and perfections of the Creator, drawn from cosmical considerations, labour under some disadvantages when compared with the arguments founded on those provisions and adaptations which more immediately affect the well-being of organized creatures. The structure of the solar system has far less analogy with such machinery as we can construct and comprehend, than we find in the structure of the bodies of animals, or even in the causes of the weather. Moreover, we do not see the immediate bearing of cosmical arrangements on that end which we most readily acknowledge to be useful and desirable, the support and comfort of sentient natures. So that, from both causes, the impression of benevolent design in this case is less striking and pointed than that which results from the examination of some other parts of nature.

But in considering the universe, according to the view we have taken, as a collection of *laws*, astronomy, the science which teaches us the laws of the motions of the heavenly bodies, possesses some advantages, among the subjects from which we may seek to learn the character of the government of the world. For our knowledge of the laws of the motions of the planets and satellites is far more complete and exact, far more thorough and satisfactory, than the knowledge which we possess in any other department of Natural Philosophy. Our acquaintance with the laws of the solar system is such, that we can calculate the precise place and motion of most of its parts at any period, past or future, however remote ; and we can refer the changes which take place in these circumstances to their proximate cause, the attraction of one mass of matter to another, acting between all the parts of the universe.

If, therefore, we trace indications of the Divine care, either in the form of the laws which prevail among the heavenly bodies, or in the arbitrary quantities which such laws involve; (according to the distinction explained in the former part of this work;) we may expect that our examples of such care, though they may be less numerous and obvious, will be more precise than they can be in other subjects, where the laws of facts are imperfectly known, and their causes entirely hid. We trust that this will be found to be the case with regard to some of the examples which we shall adduce.

Chapter I.

The Structure of the Solar System.

In the cosmical considerations which we have to offer, we shall suppose the general truths concerning the structure of the solar system and of the universe, which have been established by astronomers and mathematicians, to be known to the reader. It is not necessary to go into much detail on this subject. The five planets known to the ancients, Mercury, Venus, Mars, Jupiter, Saturn, revolve round the sun, at different distances, in orbits nearly circular, and nearly in one plane. Between Venus and Mars, our Earth, herself one of the planets, revolves in like manner. Beyond Saturn, Uranus has been discovered describing an orbit of the same kind; and between Mars and Jupiter, four smaller bodies perform their revolutions in orbits somewhat less regular than the rest. These planets are all nearly globular, and all revolve upon their axis. Some of them are accompanied by satellites, or attendant bodies which revolve about them; and these bodies also have their orbits nearly circular, and nearly in the same plane as the others. Saturn's ring is a solitary example, so far as we know, of such an appendage to a planet.

These circular motions of the planets round the sun, and of the satellites round their primary planets, are all kept going by the *attraction* of the respective central bodies, which restrains the corresponding revolving bodies from flying off. It is perhaps not very easy to make this operation clear to common apprehension. We cannot illustrate it by a comparison with any machine of human contrivance and fabrication : in such machines everything goes on by contact and impulse : pressure, and force of all kinds, is exercised and transferred from one part to another, by means of a material connexion ; by rods ropes, fluids, gases. In the machinery of the universe there is, so far as we know, no material connexion between the parts which act on each other. In the solar system no part touches or drives another : all the bodies affect each other *at a distance*, as the magnet affects the needle. The production and regulation of such effects, if attempted by our mechanicians, would require great skill and nicety of adjustment ; but our artists have not executed any examples of this sort of machinery, by reference to which we can illustrate the arrangements of the solar system.

Perhaps the following comparison may serve to explain the kind of adjustments of which we shall have to speak. If there be a wide shallow round basin of smooth marble, and if we take a smooth ball, as a billiard ball or a marble pellet, and throw it along the surface of the inside of the basin, the ball will generally make many revolutions round the inside of the bowl, gradually tending to the bottom in its motion. The gradual diminution of the motion, and consequent tendency of the ball to the bottom of the bowl, arises from the friction ; and in order to make the motion correspond to that which takes place through the action of a central force, we must suppose this friction to be got rid of. In this case, the ball, once set a going, would run round the basin for ever, describing either a circle, or various kinds of ovals, according to the way in which it was originally thrown ; whether quick or slow, and whether more or less obliquely along the surface.

Such a motion would be capable of the same kind of variety, and the same sort of adjustments, as the motion of a body revolving about a larger one by means of a central force. Perhaps the reader may understand what kind of adjustments these are, by supposing such a bowl and ball to be used for a game of skill. If the object of the players be to throw the pellet along the surface of the basin, so that after describing its curved path it shall pass through a small hole in a barrier at some distance from the starting point, it will easily be understood that some nicety in. the regulation of the force and direction with which the ball is thrown will be necessary for success. In order to obtain a better image of the solar system, we must suppose the basin to be very large and the pellet very small. And it will easily be understood that as many pellets as there are planets might run round the bowl at the same time with different velocities. Such a contrivance might form a *planetarium* in which the mimic planets would be regulated by the laws of motion as the real planets are ; instead of being carried by wires and wheels, as is done in such machines of the common construction : and in this planetarium the tendency of the planets to the sun is replaced by the tendency of the representative pellets to run down the slope of the bowl. We shall refer again to this basin, thus representing the solar system with its loose planetary balls.

Chapter II.

The Circular Orbits of the Planets round the Sun.

The orbit which the earth describes round the sun is very nearly a circle : the sun is about one thirtieth nearer to us in winter than in summer. This nearly circular form of the orbit, on a little consideration, will appear to be a remarkable circumstance.

Supposing the attraction of a planet towards the sun to exist, if the planet were put in motion in any part of the solar system, it would describe about the sun an orbit *of some kind;* it might be a long oval, or a shorter oval, or an exact circle. But if we suppose the result left to chance, the chances are infinitely against the last mentioned case. There is but one circle ; there are an infinite number of ovals. Any original impulse would give some oval, but only one particular impulse, determinate in velocity and direction, will give a circle. If we suppose the planet to be originally *projected*, it must be projected perpendicularly to its distance from the sun, and with a certain precise velocity, in order that the motion may be circular.

In the basin to which we have compared the solar system, the adjustment requisite to produce circular motion would require us to project our pellet so that after running half round the surface it should touch a point exactly at an equal distance from the centre, on the other side, passing neither too high nor too low. And the pellet, it may be observed, should be in size only one ten thousandth part of the distance from the centre, to make the dimensions correspond with the cast of the earth's orbit. If the mark were set up and hit, we should hardly attribute the result to chance.

The earth's orbit, however, is not exactly a circle. The mark is not precisely a single point, but is a space of the breadth of one thirtieth of the distance from the centre. Still this is much too near an agreement with the circle to be considered as the work of chance. The chances were great against the ball passing so nearly at the same distance, for there were twenty-nine equal spaces through which it might have gone, between the mark and the centre, and an indefinite number outside the mark.

But it is not the earth's orbit alone which is nearly a circle : the rest of the planets also approach very nearly to that form : Venus more nearly still than the earth : Jupiter, Saturn, and Uranus have a difference of about one-tenth, between their greatest and least distances from the sun : Mars has his extreme distances in the proportion of five to six nearly ; and Mercury in the proportion of two to three. The last mentioned case is a considerable deviation, and two of the

small planets which lie between Mars and
Jupiter, namely Juno and Pallas, exhibit an
inequality somewhat greater still ; but the smallness of these bodies, and other circumstances,
make it probable that there may be particular
causes for the exception in their case.   The
orbits of the satellites of the Earth, of Jupiter and
of Saturn, are also nearly circular.

Taking the solar system altogether, the regularity of its structure is very remarkable.   The
diagram which represents the orbits of the planets
might have consisted of a number of ovals, narrow and wide in all degrees, intersecting and
interfering with each other in all directions.   The
diagram does consist, as all who have opened a
book of astronomy know, of a set of figures which
appear at first sight concentric circles, and which
are very nearly so; no where approaching to any
crossing or interfering, except in the case of the
small planets, already noticed as irregular.   No
one, looking at this common diagram, can believe
that the orbits were made to be so nearly circles
by chance ; any more than he can believe that a
target, such as archers are accustomed to shoot at,
was painted in concentric circles by the accidental
dashes of a brush in the hands of a blind man.

The regularity, then, of the solar system excludes the notion of accident in the arrangement
of the orbits of the planets.   There must have
been an express adjustment to produce this cir-

cular character of the orbits. The velocity and
direction of the motion of each planet must have
been subject to some original regulation ; or, as
it is often expressed, the projectile force must
have been accommodated to the centripetal force.
This once done, the motion of each planet, taken
by itself, would go on for ever still retaining its
circular character, by the laws of motion.

If some original cause adjusted the orbits of
the planets to their circular form and regular
arrangement, we can hardly avoid including in
our conception of this cause, the intention and
will of a Creating Power. We shall consider this
argument more fully in a succeeding chapter ;
only observing here, that the presiding Intelligence, which has selected and combined the
properties of the organic creation, so that they
correspond so remarkably with the arbitrary
quantities of the system of the universe, may
readily be conceived also to have selected the
arbitrary velocity and direction of each planet's
motion, so that the adjustment should produce
a close approximation to a circular motion.

We have argued here only from the *regularity*
of the solar system ; from the selection of the
single symmetrical case and the rejection of all
the unsymmetrical cases. But this subject may
be considered in another point of view. The
system thus selected is not only regular and
symmetrical, but also it is, so far as we can

judge, the only one which would answer the purpose of the earth, perhaps of the other planets,
as the seat of animal and vegetable life.   If the
earth s orbit were more excentric, as it is called,
if for instance the greatest and least distances
were as three to one, the inequality of heat at
two seasons of the year would be destructive to
the existing species of living creatures.   A circular, or nearly circular, orbit, is the only case in
which we can have a course of seasons such as
we have at present, the only case in which the
climates of the northern and southern hemispheres
are nearly the same ; and what is more clearly
important, the only case in which the character
of the seasons would not vary from century to
century.   For if the excentricity of the earth's
orbit were considerable, the difference of heat at
different seasons, arising from the different distances of the sun, would be combined with the
difference, now the only considerable one, which
depends on the position of the earth's axis.   And
as by the motion of the *perihelion*, or place of the
nearest distance of the earth to the sun, this
nearest distance would fall in different ages at
different parts of the year, the whole distribution
of heat through the year would thus be gradually
subverted.   The summer and winter of the *tropical*
year, as we have it now, being combined with the
heat and cold of the *anomalistic* year, a period of
different length, the difference of the two seasons

might sometimes be neutralized altogether, and at other times exaggerated by the accumulation of the inequalities, so as to be intolerable.

The circular form of the orbit therefore, which, from its unique character, appears to be chosen with *some* design, from its effects on the seasons appears to be chosen with this design, so apparent in other parts of creation, of securing the welfare of organic life, by a steadfast and regular order of the solar influence upon the planet.

Chapter III.

The Stability of the Solar System.

There is a consequence resulting from the actual structure of the solar system, which has been brought to light by the investigations of mathematicians concerning the cause and laws of its motions, and which has an important bearing on our argument. It appears that the arrangement which at present obtains is precisely that which is necessary to secure the *stability* of the system. This point we must endeavour to explain.

If each planet were to revolve round the sun without being affected by the other planets, there would be a certain degree of regularity in its motion ; and this regularity would continue for ever. But it appears, by the discovery of the law of universal gravitation, that the planets do not execute their movements in this insulated and independent manner. Each of them is acted on by the attraction of all the rest. The Earth is constantly drawn by Venus, by Mars, by Jupiter, bodies of various magnitudes, perpetually changing their distances and positions with regard to the earth ; the Earth in return is perpetually drawing these bodies. What, in the course of time, will be the result of this mutual attraction ?

All the planets are very small compared with the sun, and therefore the derangement which they produce in the motion of one of their number will be very small in the course of one revolution. But this gives us no security that the derangement may not become very large in the course of many revolutions. The cause acts perpetually, and it has the whole extent of time to work in. Is it not easily conceivable then that in the lapse of ages the derangements of the motions of the planets may accumulate, the orbits may change their form, their mutual distances may be much increased or much diminished ? Is it not possible that these changes may go on without limit, and end in the complete sub-version and ruin of the system ?

If, for instance, the result of this mutual gravitation should be to increase considerably the ex-centricity of the earth's orbit, that is to make it a longer and longer oval; or to make the moon approach perpetually nearer and nearer the earth every revolution ; it is easy to see that in the one case our year would change its character, as we have noticed in the last section ; in the other, our satellite might finally fall to the earth, which must of course bring about a dreadful catastrophe. If the positions of the planetary orbits, with respect to that of the earth, were to change much, the planets might sometimes come very near us, and thus exaggerate the effects of their attraction beyond calculable limits. Under such circumstances, we might have " years of unequal length, and seasons of capricious temperature, planets and moons of portentous size and aspect, glaring and disappearing at uncertain intervals ;" tides like deluges, sweeping over whole continents , and, perhaps, the collision of two of the planets, and the consequent destruction of all organization on both of them.

Nor is it, on a common examination of the history of the solar system, at all clear that there is no tendency to indefinite derangement. The fact really is, that changes are taking place in the motions of the heavenly bodies, which have gone on progressively from the first dawn of science. The excentricity of the earth's orbit has been diminishing from the earliest observations to our times. The moon has been moving quicker and quicker from the time of the first recorded eclipses, and is now in advance, by about four times her own breadth, of what her place would have been if it had not been affected by this acceleration. The obliquity of the ecliptic also is in a state of diminution, and is now about two-fifths of a degree less than it was in the time of Aristotle. Will these changes go on without limit or reaction? If so, we tend by natural causes to a termination of the present system of things: If not, by what adjustment or combination are we secured from such a tendency? Is the system *stable*, and if so, what is the condition on which its stability depends?

To answer these questions is far from easy. The mechanical problem which they involve is no less than this;—Having given the directions and velocities with which about thirty bodies are moving at one time, to find their places and motions after any number of ages; each of the bodies, all the while, attracting all the others, and being attracted by them all.

It may readily be imagined that this is a problem of extreme complexity, when it is considered that every new *configuration* or arrangement of the bodies will give rise to a new amount of action on each; and every new action to a new configuration. Accordingly, the mathematical investigation of such questions as the above was too difficult to be attempted in the earlier periods of the progress of Physical Astronomy. Newton did not undertake to demonstrate either the stability or the instability of the system. The decision of this point required a great number of preparatory steps and simplifications, and such progress in the invention and improvement of mathematical methods, as occupied the best mathematicians of Europe for the greater part of last century. But, towards the end of that time, it was shewn by Lagrange and Laplace that the arrangements of the solar system are stable : that in the long run, the orbits and motions remain unchanged ; and that the changes in the orbits, which take place in shorter periods, never transgress certain very moderate limits. Each orbit undergoes deviations on this side and on that of its average state ; but these deviations are never very great, and it finally recovers from them, so that the average is preserved. The planets produce perpetual perturbations in each other's motions, but these perturbations are not indefinitely progressive, they are periodical : they reach a *maximum* value and then diminish. The periods which this restoration requires are, for the most part, enormous ; not less than thousands, and, in some instances, millions of years ; and hence it is, that some of these apparent derangements have been going on in the same direction since the beginning of the history of the world. But the restoration is in the sequel as complete as the derangement ; and in the meantime the disturbance never attains a sufficient amount seriously to alter the adaptations of the system.*

The same examination of the subject by which this is proved, points out also the conditions on which this stability depends. " I have succeeded in demonstrating," says Laplace, " that whatever be the masses of the planets, in consequence of the fact that they all move in the same direction, in orbits of small excentricity, and slightly inclined to each other—their secular inequalities are periodical and included within narrow limits ; so that the planetary system will only oscillate about a mean state, and will never deviate from it except by a very small quantity. The ellipses of the planets have been, and always will be, nearly circular. The ecliptic will never coincide with the equator, and the entire extent of the variation in its inclination cannot exceed three degrees."

There exists, therefore, it appears, in the solar system, a provision for the permanent regularity of its motions ; and this provision is found in the fact that the orbits of the planets are nearly circular, and nearly in the same plane, and the motions all in the same direction, namely, from west to east†.

* Laplace Expos. du Syst. du Monde. p. 441.

† In this statement of Laplace, however, one remarkable provision for the stability of the system is not noticed. The planets

Now is it probable that the occurrence of these conditions of stability in the disposition of the solar system is the work of chance? Such a supposition appears to be quite inadmissible. Any one of the orbits might have had any excentricity.* In that of Mercury, where it is much the greatest, it is only one-fifth. How came it to pass that the orbits were not more elongated? A little more or a little less velocity in their original motions would have made them so. They might have had any inclination to

Mercury and Mars, which have much the largest excentricities among the old planets, are those of which the masses are much the smallest. The mass of Jupiter is more than 2000 times that of either of these planets. If the orbit of Jupiter were as excentric as that of Mercury is, all the security for the stability of the system, which analysis has yet pointed out, would disappear. The earth and the smaller planets might in that case change their approximately circular orbits into very long ellipses, and thus might fall into the sun, or fly off into remote space.

It is further remarkable that in the newly discovered planets, of which the orbits are still more excentric than that of Mercury, the masses are still smaller, so that the same provision is established in this case also. It does not appear that any mathematician has even attempted to point out a necessary connexion between the mass of a planet and the excentricity of its orbit on any hypothesis. May we not then consider this combination of small masses with large excentricities, so important to the purposes of the world, as a mark of provident care in the Creator?

* The *excentricity* of a planet's orbit is measured by taking the proportion of the *difference* of the greatest and least distances from the sun, to the *sum* of the same distances. Mercury's greatest and least distances are as 2 and 3; his excentricity therefore is one fifth.

the ecliptic from *no* degrees to ninety degrees. Mercury, which again deviates most widely, is inclined 7¾ degrees, Venus 3¾, Saturn 2¾, Jupiter 1½, Mars 2. How came it that their motions are thus contained within such a narrow strip of the sky ? One, or any number of them, might have moved from east to west: none of them does so. And these circumstances, which appear to be, each in particular, requisite for the stability of the system and the smallness of its disturbances, are all found in combination. Does not this imply both clear purpose and profound skill ?

It is difficult to convey an adequate notion of the extreme complexity of the task thus executed. A number of bodies, all attracting each other, are to be projected in such a manner that their revolutions shall be permanent and stable, their mutual perturbations always small. If we return to the basin with its rolling balls, by which we before represented the solar system, we must complicate with new conditions the trial of skill which we supposed. The problem must now be to project at once seven such balls, all connected by strings which influence their movements, so that each may hit its respective mark. And we must further suppose, that the marks are to be hit after many thousand revolutions of the balls. No one will imagine that this could be done by accident.

In fact it is allowed by all those who have

considered this subject, that such a coincidence of the existing state with the mechanical requisites of permanency cannot be accidental. Laplace has attempted to calculate the probability that it is not the result of accident. He takes into account, in addition to the motions which we have mentioned, the revolutions of the satellites about their primaries, and of the sun and planets about their axes : and he finds that there is a probability, far higher than that which we have for the greater part of undoubted historical events, that these appearances are not the effect of chance. "We ought, therefore," he says, "to believe, with at least the same confidence, that a primitive cause has directed the planetary motions."

The solar system is thus, by the confession of all sides, completely different from anything which we might anticipate from the casual operation of its known laws. The laws of motion are no less obeyed to the letter in the most irregular than in the most regular motions ; no less in the varied circuit of the ball which flies round a tennis court, than in the going of a clock ; no less in the fantastical jets and leaps which breakers make when they burst in a corner of a rocky shore, than in the steady swell of the open sea. The laws of motion alone will not produce the regularity which we admire in the motions of the heavenly bodies. There must be an original adjustment of the system on which these laws are to act ; a selection of the arbitrary quantities which they are to involve ; a primitive cause which shall dispose the elements in due relation to each other, in order that regular recurrence may accompany constant change ; that perpetual motion may be combined with perpetual stability ; that derangements which go on increasing for thousands or for millions of years may finally cure themselves ; and that the same laws which lead the planets slightly aside from their paths, may narrowly limit their deviations, and bring them back from their almost imperceptible wanderings.

If a man does not deny that any possible peculiarity in the disposition of the planets with regard to the sun could afford evidence of a controlling and ordering purpose, it seems difficult to imagine how he could look for evidence stronger than that which there actually is.   Of all the innumerable possible cases of systems, governed by the existing laws of force and motion, that one is selected which alone produces such a steadfast periodicity, such a constant average of circumstances, as are, so far as we can conceive, necessary conditions for the existence of organic and sentient life. And this selection is so far from being an obvious or easily discovered means to this end, that the most profound and attentive consideration of the properties of space and number, with all the appliances and aids we can obtain, are barely sufficient to enable *us* to see that the end is thus secured, and that it can be secured in no other way. Surely the obvious impression which arises from this view of the subject is, that the solar system, with its adjustments, is the work of an Intelligence, who perceives, as self evident, those truths, to which we attain painfully and slowly, and after all imperfectly ; who has employed in every part of creation refined contrivances, which we can only with effort understand ; and who, in innumerable instances, exhibits to us what we should look upon as remarkable difficulties re-markably overcome, if it were not that, through the perfection of the provision, the trace of the difficulty is almost obliterated.

CHAPTER IV.

The Sun in the Centre.

THE next circumstance which we shall notice as indicative of design in the arrangement of the material portions of the solar system, is the posi-tion of the sun, the source of light and heat, in the centre of the system. This could hardly have occurred by any thing which we can call chance. Let it be granted, that the law of gra-vitation is established, and that we have a large mass, with others much smaller in its comparative vicinity. The small bodies may then move round the larger, but this will do nothing towards making it a *sun* to them. Their motions might take place, the whole system remaining still utterly dark and cold, without day or summer. In order that we may have something more than this blank and dead assemblage of moving clods, the machine must be lighted up and warmed. Some of the advantages of placing the lighting and warming apparatus in the centre are obvious to us. It is in this way only that we could have those regular periodical returns of solar influence, which, as we have seen, are adapted to the constitution of the living creation. And we can easily conceive, that there may be other incongruities in a system with a travelling sun, of which we can only conjecture the nature. No one probably will doubt that the existing system, with the sun in the centre, is better than any one of a different kind would be.

Now this lighting and warming by a central sun are something superadded to the mere mechanical arrangements of the universe. There is no apparent reason why the largest mass of gravitating matter should diffuse inexhaustible supplies of light and heat in all directions, while the other masses are merely passive, with respect to such influences. There is no obvious connexion between mass and luminousness, or temperature. No one, probably, will contend that the materials of our system are necessarily luminous or hot. According to the conjectures of astronomers, the heat and light of the sun do not reside in its mass, but in a coating which lies on its surface. If such a coating were fixed there by the force of universal gravitation, how could we avoid having a similar coating on the surface of the earth, and of all the other globes of the system. If light consist in the vibrations of an ether, which we have mentioned as a probable opinion, why has the sun alone the power of exciting such vibrations? If light be the emission of material particles, why does the sun alone emit such particles? Similar questions may be asked, with regard to heat, whatever be the theory we adopt on that subject. Here then we appear to find marks of contrivance. The sun might become, we will suppose, the centre of the motions of the planets by mere mechanical causes: but what caused the centre of their motions to be also the source of those vivifying influences? Allowing that no interposition was requisite to regulate the revolutions of the system, yet observe what a peculiar arrangement in other respects was necessary, in order that these revolutions might produce days and seasons! The machine will move of itself, we may grant: but who constructed the machine, so that its movements might answer the purposes of life? How

was the candle placed upon the candlestick ?
how was the fire deposited on the hearth, so that
the comfort and well-being of the family might
be secured ?   Did these too fall into their places
by the casual operation of gravity ? and, if not,
is there not here a clear evidence of intelligent
design, of arrangement with a benevolent end ?

This argument is urged with great force by
Newton himself.   In his first letter to Bentley,
he allows that matter might form itself into
masses by the force of attraction.   " And thus,"
says he, " might the sun and fixed stars be formed,
supposing the matter were of a lucid nature.
But how the matter should divide itself into two
sorts ; and that part of it which is fit to compose
a shining body should fall down into one mass,
and make a sun ; and the rest, which is fit to
compose an opake body, should coalesce, not
into one great body, like the shining matter, but
into many little ones ; or if the sun at first were
an opake body like the planets, or the planets
lucid bodies like the sun, how he alone should be
changed into a shining body, whilst all they
continue opake; or all they be changed into opake
ones, while he continued unchanged : I do not
think explicable by mere natural causes, but am
forced to ascribe it to the counsel and contrivance
of a voluntary Agent."

Chapter V.

The Satellites.

1. A person of ordinary feelings, who, on a fine moonlight night, sees our satellite pouring her mild radiance on field and town, path and moor, will probably not only be disposed to " bless the useful light," but also to believe that it was " ordained" for that purpose ;—that the lesser light was made to rule the night as certainly as the greater light was made to rule the day.

Laplace, however, does not assent to this belief. He observes, that " some partisans of final causes have imagined that the moon was given to the earth to afford light during the night :" but he remarks that this cannot be so, for that we are often deprived at the same time of the light of the sun and the moon ; and he points out how the moon might have been placed so as to be always " full."

That the light of the moon affords, *to a certain extent*, a supplement to the light of the sun, will hardly be denied. If we take man in a condition in which he uses artificial light scantily only, or not at all, there can be no doubt that the moonlight nights are for him a very important addition to the time of daylight. And as a small proportion only of the whole number of nights are without some portion of moonlight, the fact that sometimes both luminaries are invisible very little diminishes the value of this advantage. Why we have not more moonlight, either in duration or in quantity, is an enquiry which a philosopher could hardly be tempted to enter upon, by any success which has attended previous speculations of a similar nature. Why should not the moon be ten times as large as she is? Why should not the pupil of man s eye be ten times as large as it is, so as to receive more of the light which does arrive? We do not conceive that our inability to answer the latter question prevents our knowing that the eye was made for seeing: nor does our inability to answer the former, disturb our persuasion that the moon was made to give light upon the earth.

Laplace suggests that if the moon had been placed at a certain distance beyond the earth, it would have revolved about the sun in the same time as the earth does, and would have always presented to us a full moon. For this purpose it must have been about four times as far from us as it really is; and would therefore, other things remaining unchanged, have only been *one sixteenth* as large to the eye as our present full moon. We shall not dwell on the discussion of this suggestion, for the reason just intimated. But we may observe that in such a system as

Laplace proposes, it is not yet proved, we believe, that the arrangement would be stable under the influence of the disturbing forces. And we may add that such an arrangement, in which the motion of one body has a *coordinate* reference to two others, as the motion of the moon on this hypothesis would have to the sun and the earth, neither motion being subordinate to the other, is contrary to the whole known analogy of cosmical phenomena, and therefore has no claim to our notice as a subject of discussion.

2. In turning our consideration to the satellites of the other planets of our system, there is one fact which immediately arrests our attention;— the number of such attendant bodies appears to increase as we proceed to planets farther and farther from the sun. Such at least is the general rule. Mercury and Venus, the planets nearest the sun, have no such attendants : the Earth has one. Mars, indeed, who is still farther removed, has none ; nor have the minor planets, Juno, Vesta, Ceres, Pallas ; so that the rule is only approximately verified. But Jupiter, who is at five times the earth's distance, has four satellites ; and Saturn, who is again at a distance nearly twice as great, has seven, besides that most extraordinary phenomenon his ring, which, for purposes of illumination, is equivalent to many thousand satellites. Of Uranus it is difficult to speak, for his great distance renders it almost

impossible to observe the smaller circumstances of his condition. It does not appear at all probable that he has a ring, like Saturn; but he has at least five satellites which are visible to us, at the enormous distance of 900 millions of miles ; and we believe that the astronomer will hardly deny that he may possibly have thousands of smaller ones circulating about him.

But leaving conjecture, and taking only the ascertained cases of Venus, the Earth, Jupiter, and Saturn, we conceive that a person of common understanding will be strongly impressed with the persuasion that the satellites are placed in the system with a view to compensate for the diminished light of the sun at greater distances. The smaller planets, Juno, Vesta, Ceres, and Pallas, differ from the rest in so many ways, and suggest so many conjectures of reasons for such differences, that we should almost expect to find them exceptions to such a rule. Mars is a more obvious exception. Some persons might conjecture from his case, that the arrangement itself, like other useful arrangements, has been brought about by some wider law which we have not yet detected. But whether or not we entertain such a guess, (it can be nothing more,) we see in other parts of creation, so many examples of apparent exceptions to rules, which are afterwards found to be explained, or provided for by particular contrivances, that no one, familiar with such contemplations, will, by one anomaly, be driven from the persuasion that the end which the arrangements of the satellites seem suited to answer is really one of the ends of their creation.

CHAPTER VI.

The Stability of the Ocean.

WHAT is meant by the stability of the ocean may perhaps be explained by means of the following illustration. If we suppose the whole globe of the Earth to be composed of water, a sphere of cork, immersed in any part of it, would come to the surface of the water, except it were placed exactly at the centre of the earth; and even if it were the slightest displacement of the cork sphere would end in its rising and floating. This would be the case whatever were the size of the cork sphere, and even if it were so large as to leave comparatively little room for the water; and the result would be nearly the same, if the cork sphere, when in its central position, had on its surface prominences which projected above the surface of the water. Now this brings us to the case in which we have a globe resembling our present earth, composed like it of water and of a solid centre, with islands and continents, but having these solid parts all made of cork. And it appears by the preceding reasoning, that in this case, if there were any disturbance either of the solid or fluid parts, the solid parts would rise from the centre of the watery sphere as far as they could : that is, all the water would run to one side and leave the land on the other. Such an ocean would be in *unstable* equilibrium.

Now a question naturally occurs, is the equilibrium of our present ocean of this unstable kind, or is it stable? The sea, after its most violent agitations, appears to return to its former state of repose ; but may not some extraordinary cause produce in it some derangement which may go on increasing till the waters all rush one way, and thus drown the highest mountains? And if we are safe from this danger, what are the conditions by which we are so secured ?

The illustration which we have employed obviously suggests the answer to this question ; namely, that the equilibrium is unstable, so long as the solid parts are of such a kind as to float in the fluid parts ; and of course we should expect that the equilibrium will be stable whenever the contrary is the case, that is, when the solid parts of the earth are of greater specific gravity than the sea. A more systematic mathematical calculation has conducted Laplace to a demonstration of this result.

The mean specific gravity of the earth appears to be about *five* times that of water, so that the condition of the stability of the ocean is abundantly fulfilled. And the provision by which this stability is secured was put in force through the action of those causes, whatever they were, which made the density of the solid materials and central parts of the earth greater than the density of the incumbent fluid.

When we consider, however, the manner in which the wisdom of the Creator, even in those cases in which his care is most apparent, as in the structure of animals, works by means of intermediate causes and general laws, we shall not be ready to reject all belief of an end in such a case as this, merely because the means are mechanical agencies. Laplace says, " in virtue of gravity, the most dense of the strata of the earth are those nearest to the centre ; and thus the mean density exceeds that of the waters which cover it ; which suffices to secure the stability of the equilibrium of the seas, and to put a bridle upon the fury of the waves." This statement, if exact, would not prove that He who subjected the materials of the earth to the action of gravity did not *intend* to restrain the rage of the waters : but the statement is not true in fact. The lower strata, so far as man has yet examined, are very far from being constantly, or even generally, heavier than the superincumbent ones. And certainly solidification by no means implies a greater density than fluidity : the density of

Jupiter is one fourth, that of Saturn less than one seventh, of that of the earth. If an ocean of water were poured into the cavities upon the surface of Saturn, its equilibrium would *not* be stable. It would leave its bed on one side of the globe ; and the planet would finally be composed of one hemisphere of water and one of land. If the Earth had an ocean of a fluid six times as heavy as water, (quicksilver is thirteen times as heavy,) we should have, in like manner, a dry and a fluid hemisphere. Our inland rivers would probably never be able to reach the shores, but would be dried up on their way, like those which run in torrid desarts ; perhaps the evaporation from the ocean would never reach the inland mountains, and we should have no rivers at all. Without attempting to imagine the details of such a condition, it is easy to see, that to secure the existence of a different one is an end which is in harmony with all that we see of the preserving care displayed in the rest of creation.*

---

* The stability of the axis of rotation about which the earth revolves has sometimes been adduced as an instance of preservative care. This stability, however, would follow necessarily, if the earth, or its superficial parts, were originally fluid ; and that they were so is an opinion widely received, both among astronomers and geologists. The original fluidity of the earth is probably a circumstance depending upon the general scheme of creation ; and cannot with propriety be considered with reference to one particular result. We shall therefore omit any further consideration of this argument.

Chapter VII.

The Nebular Hypothesis.

WE have referred to Laplace, as a profound mathematician, who has strongly expressed the opinion, that the arrangement by which the stability of the solar system is secured is not the result of chance; that " *a primitive cause* has directed the planetary motions." This author, however, having arrived, as we have done, at this conviction, does not draw from it the conclusion which has appeared to us so irresistible, that " the admirable arrangement of the solar system cannot but be the work of an intelligent and most powerful being." He quotes these expressions, which are those of Newton, and points at them as instances where that great philosopher had deviated from the method of true philosophy. He himself proposes an hypothesis concerning the nature of the *primitive cause* of which he conceives the existence to be thus probable : and this hypothesis, on account of the facts which it attempts to combine, the view of the universe which it presents, and the eminence of the person by whom it is propounded, deserves our notice.

1. Laplace conjectures that in the original condition of the solar system, the sun revolved upon

his axis, surrounded by an atmosphere which, in
virtue of an excessive heat, extended far beyond
the orbits of all the planets, the planets as yet
having no existence. The heat gradually diminished, and as the solar atmosphere contracted
by cooling, the rapidity of its rotation increased
by the laws of rotatory motion, and an exterior
zone of vapour was detached from the rest, the
central attraction being no longer able to overcome the increased centrifugal force. This zone
of vapour might in some cases retain its form, as
we see it in Saturn's ring ; but more usually the
ring of vapour would break into several masses,
and these would generally coalesce into one
mass, which would revolve about the sun. Such
portions of the solar atmosphere, abandoned
successively at different distances, would form
" planets in the state of vapour." These planets,
it appears from mechanical considerations, would
have each its rotatory motion, and as the cooling
of the vapour still went on, would each produce
a planet, which might have satellites and rings,
formed from the planet in the same manner as
the planets were formed from the atmosphere of
the sun.

It may easily be conceived that all the primary
motions of a system so produced would be nearly
circular, nearly in the plane of the original
equator of the solar rotation, and in the direction
of that rotation. Reasons are offered also to

shew that the motions of the satellites thus produced and the motions of rotation of the planets
must be in the same direction. And thus it is
held that the hypothesis accounts for the most
remarkable circumstances in the structure of the
solar system : namely, the motions of the planets in
the same direction, and almost in the same plane ;
the motions of the satellites in the same direction
as those of the planets ; the motions of rotation of
these different bodies still in the same direction
as the other motions, and in planes not much
different ; the small excentricity of the orbits of
the planets, upon which condition, along with
some of the preceding ones, the stability of the
system depends ; and the position of the source
of light and heat in the centre of the system.

It is not necessary for the purpose, nor suitable
to the plan of the present treatise, to examine,
on physical grounds, the probability of the above
hypothesis. It is proposed by its author, with
great diffidence, as a conjecture only. We might,
therefore, very reasonably put off all discussion
of the bearings of this opinion upon our views of
the government of the world, till the opinion
itself should have assumed a less indistinct and
precarious form. It can be no charge against
our doctrines, that there is a difficulty in reconciling with them arbitrary guesses and half-
formed theories. We shall, however, make a

few observations upon this *nebular hypothesis*, as it may be termed.

2. If we grant, for a moment, the hypothesis, it by no means proves that the solar system was formed without the intervention of intelligence and design. It only transfers our view of the skill exercised, and the means employed, to another part of the work. For, how came the sun and its atmosphere to have such materials, such motions, such a constitution, that these consequences followed from their primordial condition? How came the parent vapour thus to be capable of coherence, separation, contraction, solidification? How came the laws of its motion, attraction, repulsion, condensation, to be so fixed, as to lead to a beautiful and harmonious system in the end? How came it to be neither too fluid nor too tenacious, to contract neither too quickly nor too slowly, for the successive formation of the several planetary bodies? How came that substance, which at one time was a luminous vapour, to be at a subsequent period, solids and fluids of many various kinds? What but design and intelligence prepared and tempered this previously existing element, so that it should by its natural changes produce such an orderly system?

And if in this way we suppose a planet to be produced, what sort of a body would it be?— something, it may be presumed, resembling a large meteoric stone. How comes this mass to be covered with motion and organization, with life and happiness? What primitive cause stocked it with plants and animals, and produced all the wonderful and subtle contrivances which we find in their structure, all the wide and profound mutual dependencies which we trace in their economy? Was man, with his thought and feeling, his powers and hopes, his will and conscience, also produced as an ultimate result of the condensation of the solar atmosphere? Except we allow a prior purpose and intelligence presiding over this material " primitive cause," how irreconcilable is it with the evidence which crowds in upon us from every side!

3. In the next place, we may observe concerning this hypothesis, that it carries us back to the beginning of the present system of things; but that it is impossible for our reason to stop at the point thus presented to it. The sun, the earth, the planets, the moons were brought into their present order out of a previous state, and, as is supposed in the theory, by the natural operation of laws. But how came that previous state to exist? We are compelled to suppose that it, in like manner, was educed from a still prior state of things; and this, again, must have been the result of a condition prior still. Nor is it possible for us to find, in the tenets of the nebular hypothesis, any resting place or satisfaction for

the mind. The same reasoning faculty, which seeks for the origin of the present system of things, and is capable of assenting to, or dissenting from the hypothesis propounded by Laplace as an answer to this enquiry, is necessarily led to seek, in the same manner, for the origin of any previous system of things, out of which the present may appear to have grown : and must pursue this train of enquiries unremittingly, so long as the answer which it receives describes a mere assemblage of matter and motion ; since it would be to contradict the laws of matter and the nature of motion, to suppose such an assemblage to be the *first* condition.

The reflection just stated, may be illustrated by the further consideration of the Nebular Hypothesis. This opinion refers us, for the origin of the solar system, to a sun surrounded with an atmosphere of enormously elevated temperature, revolving and cooling. But as we ascend to a still earlier period, what state of things are we to suppose?—a still higher temperature, a still more diffused atmosphere. Laplace conceives that, in its primitive state, the sun consisted in a diffused luminosity so as to resemble those nebulæ among the fixed stars, which are seen by the aid of the telescope, and which exhibit a nucleus, more or less brilliant, surrounded by a cloudy brightness. " This anterior state was itself preceded by other states, in which the nebulous matter was more and more diffuse, the nucleus being less and less luminous. We arrive," Laplace says, " in this manner, at a nebulosity so diffuse, that its existence could scarcely be suspected."

" Such is," he adds, " in fact, the first state of the nebulæ which Herschel carefully observed by means of his powerful telescopes. He traced the progress of condensation, not indeed on one nebula, for this progress can only become perceptible to us in the course of centuries ; but in the assemblage of nebulæ ; much in the same manner as in a large forest we may trace the growth of trees among the examples of different ages which stand side by side. He saw in the first place the nebulous matter dispersed in patches, in the different parts of the sky. He saw in some of these patches this matter feebly condensed round one or more faint nuclei. In other nebulæ, these nuclei were brighter in proportion to the surrounding nebulosity ; when by a further condensation the atmosphere of each nucleus becomes separate from the others, the result is multiple nebulous stars, formed by brilliant nuclei very near each other, and each surrounded by an atmosphere : sometimes the nebulous matter condensing in a uniform manner has produced nebulous systems which are called *planetary*. Finally, a still greater degree of condensation transforms all these nebulous systems into stars. The nebulæ, classed according to this philosophical view, indicate with extreme probability their future transformation into stars, and the anterior nebulous condition of the stars which now exist."

It appears then that the highest point to which this series of conjectures can conduct us, is " an extremely diffused nebulosity," attended, we may suppose, by a far higher degree of heat, than that which, at a later period of the hypothetical process, keeps all the materials of our earth and planets in a state of vapour. Now is it not impossible to avoid asking, whence was this light, this heat, this diffusion ? How came the laws which such a state implies, to be already in existence ? Whether light and heat produce their effects by means of fluid vehicles or otherwise, they have complex and varied laws which indicate the existence of some subtle machinery for their action. When and how was this machinery constructed ? Whence too that enormous expansive power which the nebulous matter is supposed to possess ? And if, as would seem to be supposed in this doctrine, all the material ingredients of the earth existed in this diffuse nebulosity, either in the state of vapour, or in some state of still greater expansion, whence were they and their properties ? how came there to be of each simple substance which now enters into the composition of the universe, just so much

and no more ?   Do we not, far more than ever,
require an origin of this origin ? an explanation
of this explanation ?   Whatever may be the
merits of the opinion as a physical hypothesis,
with which we do not here meddle, can it for a
moment prevent our looking beyond the hypothesis, to a First Cause, an Intelligent Author,
an origin proceeding from free volition, not from
material necessity ?

But again : let us ascend to the highest point
of the hypothetical progression : let us suppose
the nebulosity diffused throughout all space, so
that its course of running into patches is not yet
begun.   How are we to suppose it distributed ?
Is it equably diffused in every part ? clearly
not ; for if it were, what should cause it to
gather into masses, so various in size, form and
arrangement ?   The separation of the nebulous
matter into distinct nebulæ implies necessarily
some original inequality of distribution ; some
determining circumstances in its primitive condition.   Whence were these circumstances ? this
inequality ? we are still compelled to seek some
ulterior agency and power.

Why must the primeval condition be one of
change at all ?   Why should not the nebulous
matter be equably diffused throughout space,
and continue for ever in its state of equable
diffusion, as it must do, from the absence of all
cause to determine the time and manner of its

separation ? why should this nebulous matter grow cooler and cooler ? why should it not retain for ever the same degree of heat, whatever heat be ? If heat be a fluid, if to cool be to part with this fluid, as many philosophers suppose, what becomes of the fluid heat of the nebulous matter, as the matter cools down ? Into what unoccupied region does it find its way ?

Innumerable questions of the same kind might be asked, and the conclusion to be drawn is, that every new physical theory which we include in our view of the universe, involves us in new difficulties and perplexities, if we try to erect it into an ultimate and final account of the existence and arrangement of the world in which we live. With the evidence of such theories, considered as scientific generalizations of ascertained facts, with their claims to a place in our natural philosophy, we have here nothing to do. But if they are put forwards as a disclosure of the ultimate cause of that which occurs, and as superseding the necessity of looking further or higher ; if they claim a place in our Natural Theology, as well as our Natural Philosophy ; we conceive that their pretentions will not bear a moment's examination.

Leaving then to other persons and to future ages to decide upon the scientific merits of the nebular hypothesis, we conceive that the final fate of this opinion can not, in sound reason, affect at all the view which we have been endeavouring to illustrate;—the view of the universe as the work of a wise and good Creator. Let it be supposed that the point to which this hypothesis leads us, is the ultimate point of physical science : that the farthest glimpse we can obtain of the material universe by our natural faculties, shews it to us occupied by a boundless abyss of luminous matter : still we ask, how space came to be thus occupied, how matter came to be thus luminous ? If we establish by physical proofs, that the first fact which can be traced in the history of the world, is that "there was light;" we shall still be led, even by our natural reason, to suppose that before this could occur, " God said, let there be light."

Chapter VIII.

The Existence of a Resisting Medium in the Solar System.

THE question of a *plenum* and a *vacuum* was formerly much debated among those who speculated concerning the constitution of the universe ; that is, they disputed whether the celestial and terrestrial spaces are absolutely full, each portion being occupied by some matter or other ; or whether there are, between and among the material parts of the world, empty spaces free from all matter, however rare. This question was

often treated by means of abstract conceptions
and *à priori* reasonings ; and was sometimes considered as one in which the result of the struggle
between rival systems of philosophy, the Car-
tesian and Newtonian for instance, was involved.
It was conceived by some that the Newtonian
doctrine of the motions of the heavenly bodies,
according to mechanical laws, required that the
space in which they moved should be, absolutely
and metaphysically speaking, a vacuum.

This, however, is not necessary to the truth of the
Newtonian doctrines, and does not appear to have
been intended to be asserted by Newton himself.
Undoubtedly, according to his theory, the motions
of the heavenly bodies were calculated *on the supposition* that they do move in a space void of
any resisting fluid ; and the comparison of the
places so calculated with the places actually observed, (continued for a long course of years, and
tried in innumerable cases,) did not shew any
difference which implied the existence of a resisting fluid.   The Newtonian, therefore, was
justified in asserting that *either* there was no such
fluid, *or* that it was so thin and rarefied, that no
phenomenon yet examined by astronomers was
capable of betraying its effects.

This was all that the Newtonian needed or
ought to maintain ; for his philosophy, founded
altogether upon observation, had nothing to do
with abstract possibilities and metaphysical ne-

cessities   And in the same manner in which observation and calculation thus showed that there could be none but a very rare medium pervading the solar system, it was left open to observation and calculation to prove that there was such a medium, if any facts could be discovered which offered suitable evidence.

Within the last few years, facts have been observed which show, in the opinion of some of the best mathematicians of Europe, that such a very rare medium does really occupy the spaces in which the planets move ; and it may be proper and interesting to consider the bearing of this opinion upon the views and arguments which we have had here to present.

1. Reasons might be offered, founded on the universal diffusion of light and on other grounds, for believing that the planetary spaces cannot be entirely free from matter of some kind ; and wherever matter is, we should expect resistance. But the facts which have thus led astronomers to the conviction that such a resisting medium really exists, are certain circumstances occurring in the motion of a body revolving round the sun, which is now usually called *Encke's comet*. This body revolves in a very excentric or oblong orbit, its greatest or aphelion distance from the sun, and its nearest or perihelion distance, being in the proportion of more than ten to one. In this respect it agrees with other comets ; but its time of revolution

about the sun is much less than that of the comets which have excited most notice; for while they appear only at long intervals of years, the body of which we are now speaking returns to its perihelion every 1208 days, or in about three years and one-third. Another observable circumstance in this singular body, is its extreme apparent tenuity: it appears as a loose indefinitely formed speck of vapour, through which the stars are visible with no perceptible diminution of their brightness. This body was first seen by Mechain and Messier, in 1786,* but they obtained only two observations, whereas three, at least, are requisite to determine the path of a heavenly body. Miss Herschel discovered it again in 1795, and it was observed by several European astronomers. In 1805 it was again seen, and again in 1819. Hitherto it was supposed that the four comets thus observed were all different; Encke, however, showed that the observations could only be explained by considering them as returns of the same revolving body; and by doing this, well merited that his name should be associated with the subject of his discovery. The return of this body in 1822, was calculated beforehand, and observed in New South Wales, the comet being then in the southern part of the heavens; but on comparing the calculated and

* Airy on Encke's Comet, p. 1. note.

the observed places, Encke concluded that the observations could not be exactly explained, without supposing a resisting medium. This comet was again generally observed in Europe in 1825 and 1828, and the circumstances of the last appearance were particularly favourable for determining the absolute amount of the retardation arising from the medium, which the other observations had left undetermined.

The effect of this retarding influence is, as might be supposed from what has already been said, extremely slight; and would probably not have been perceptible at all, but for the loose texture and small quantity of matter of the revolving body. It will easily be conceived that a body which has perhaps no more solidity or coherence than a cloud of dust, or a wreath of smoke, will have less force to make its way through a fluid medium, however thin, than a more dense and compact body would have. In atmospheric air much rarefied, a bullet might proceed for miles without losing any of its velocity, while such a loose mass as the comet is supposed to be would lose its projectile motion in the space of a few yards. This consideration will account for the circumstance, that the existence of such a medium has been detected by observing the motions of Encke's comet, though the motions of the heavenly bodies previously observed showed no trace of such an impediment.

It will perhaps appear remarkable that a body so light and loose as we have described this comet to be, should revolve about the sun by laws as fixed and certain as those which regulate the motions of those great and solid masses, the Earth and Jupiter. It is however certain from observation, that this comet is acted upon by exactly the same force of solar attraction, as the other bodies of the system ; and not only so, but that it also experiences the same kind of disturbing force from the action of the other planets, which they exercise upon each other. The effect of all these causes has been calculated with great care and labour ; and the result has been an agreement with observation sufficiently close to show that these causes really act, but at the same time a *residual phenomenon* (as Sir J. Herschel expresses it) has come to light : and from this has been collected the inference of a resisting medium.

This medium produces a very small effect upon the motion of the comet, as will easily be supposed from what has been said. By Encke's calculation, it appears that the effect of the resistance, supposing the comet to move in the earth's orbit, would be about 1-850th of the sun's force on the body. The effect of such a resistance may appear, at first sight, paradoxical ; it would be to make the comet *move* more slowly, but *perform its revolutions* more quickly. This, how-

ever, will perhaps be understood if it be considered that by *moving* more slowly the comet
will be more rapidly *drawn* towards the centre,
and that in this way a revolution will be described
by a shorter path than it was before. It appears,
that in getting round the sun, the comet gains
more in this way than it loses by the diminution
of its velocity. The case is much like that of a
stone thrown in the air; the stone moves more
slowly than it would do if there were no air; but
yet it comes to the earth *sooner* than it would do
on that supposition.

It appears that the effect of the resistance of
the etherial medium, from the first discovery of
the comet up to the present time, has been to
diminish the time of revolution by about two days:
and the comet is ten days in advance of the place
which it would have reached, if there had been
no resistance.

2. The same medium which is thus shown to
produce an effect upon Encke's comet, must also
act upon the planets which move through the
same spaces. The effect upon the planets, however, must be very much smaller than the effect
upon the comet, in consequence of their greater
quantity of matter.

It is not easy to assign any probable value, or
even any certain limit, to the effect of the resisting medium upon the planets. We are entirely
ignorant of the comparative mass of the comet,

and of any of the planets ; and hence, cannot make any calculation founded on such a comparison.   Newton has endeavoured to show how small the resistance of the medium must be, if it exist.*   The result of his calculation is, that if we take the density of the medium to be that which our air will have at 200 miles from the earth's surface, supposing the law of diminution of density to go on unaltered, and if we suppose Jupiter to move in such a medium, he would in a million years lose less than a millionth part of his velocity. If a planet, revolving about the sun, were to lose any portion of its velocity by the effect of resistance, it would be drawn proportionally nearer the sun, the tendency towards the centre being no longer sufficiently counteracted by that centrifugal force which arises from the body's velocity. And if the resistance were to continue to act, the body would be drawn perpetually nearer and nearer to the centre, and would describe its revolutions quicker and quicker, till at last it would reach the central body, and the system would cease to be a system.

This result is true, however small be the velocity lost by resistance ; the only difference being, that when the resistance is small, the time requisite to extinguish the whole motion will be proportionally longer.   In all cases the times

* Principia, b. iii. prop. x.

which come under our consideration in problems
of this kind, are enormous to common apprehension. Thus Encke's comet, according to the
results of the observations already made, will
lose, in ten revolutions, or thirty-three years,
less than 1-1000th of its velocity : and if
this law were to continue, the velocity would
not be reduced to one-half its present value in
less than seven thousand revolutions or twenty-
three thousand years. If Jupiter were to lose
one-millionth of his velocity in a million years,
(which, as has been seen, is far more than
can be considered in any way probable), he
would require seventy millions of years to lose
1-1000th of the velocity ; and a period seven
hundred times as long to reduce the velocity to
one-half. These are periods of time which quite
overwhelm the imagination ; and it is not pretended that the calculations are made with any
pretensions to accuracy. But at the same time
it is beyond doubt that though the intervals of
time thus assigned to these changes are highly
vague and uncertain, the changes themselves
must, sooner or later, take place, in consequence
of the existence of the resisting medium. Since
there is such a retarding force perpetually acting,
however slight it be, it must in the end destroy
all the celestial motions. It may be millions of
millions of years before the earth's retardation
may perceptibly affect the apparent motion of

the sun ; but still the day will come (if the same
Providence which formed the system, should
permit it to continue so long) when this cause
will entirely change the length of our year and
the course of our seasons, and finally stop the
earth's motion round the sun altogether. The
smallness of the resistance, however small we
choose to suppose it, does not allow us to escape
this certainty. There is a resisting medium ;
and, therefore, the movements of the solar system
cannot go on for ever. The moment such a fluid
is ascertained to exist, the eternity of the movements of the planets becomes as impossible as a
perpetual motion on the earth.

3. The vast periods which are brought under
our consideration in tracing the effects of the
resisting medium, harmonize with all that we
learn of the constitution of the universe from
other sources. Millions, and millions of millions
of years are expressions that at first sight appear
fitted only to overwhelm and confound all our
powers of thought ; and such numbers are, no
doubt beyond the limits of any thing which we
distinctly conceive. But our powers of conception are suited rather to the wants and uses of
common life, than to a complete survey of the
universe. It is in no way unlikely that the
whole duration of the solar system should be a
period immeasurably great in our eyes, though
demonstrably finite. Such enormous numbers
have been brought under our notice by all the

advances we have made in our knowledge of nature. The smallness of the objects detected by the microscope and of their parts ; – the multitude of the stars which the best telescopes of modern times have discovered in the sky ;—the duration assigned to the globe of the earth by geological investigation ;—all these results require for their probable expression, numbers, which so far as we see, are on the same gigantic scale as the number of years in which the solar system will become entirely deranged. Such calculations depend in some degree on our relation to the vast aggregate of the works of our Creator ; and no person who is accustomed to meditate on these subjects will be surprised that the numbers which such an occasion requires should oppress our comprehension. No one who has dwelt on the thought of a universal Creator and Preserver, will be surprised to find the conviction forced upon the mind by every new train of speculation, that viewed in reference to Him, our space is a point, our time a moment, our millions a handful, our permanence a quick decay.

Our knowledge of the vast periods, both geological and astronomical, of which we have spoken, is most slight. It is in fact little more than that such periods exist ; that the surface of the earth has, at wide intervals of time, undergone great changes in the disposition of land and water, and in the forms of animal life ; and that the motions of the heavenly bodies round the sun are affected, though with inconceivable slowness, by a force which must end by deranging them altogether. It would therefore be rash to endeavour to establish any analogy between the periods thus disclosed; but we may observe that they *agree* in this, that they reduce all things to the general rule of *finite duration*. As all the geological states of which we find evidence in the present state of the earth, have had their termination, so also the astronomical conditions under which the revolutions of the earth itself proceed, involve the necessity of a future cessation of these revolutions.

The contemplative person may well be struck by this universal law of the creation. We are in the habit sometimes of contrasting the transient destiny of man with the permanence of the forests, the mountains, the ocean,—with the unwearied circuit of the sun. But this contrast is a delusion of our own imagination: the difference is after all but one of degree. The forest tree endures for its centuries and then decays; the mountains crumble and change, and perhaps subside in some convulsion of nature; the sea retires, and the shore ceases to resound with the 'everlasting' voice of the ocean: such reflexions have already crowded upon the mind of the geologist; and it now appears that the courses of the heavens themselves are not exempt from the universal law of decay; that not only the rocks and the mountains, but the sun and the moon have the sentence " to end" stamped upon their foreheads.   They enjoy no privilege beyond man except a longer respite.   The ephemeron perishes in an hour ; man endures for his three score years and ten ; an empire, a nation, numbers its centuries, it may be its thousands of years ; the continents and islands which its dominion includes have perhaps their date, as those which preceded them have had ; and the very revolutions of the sky by which centuries are numbered will at last languish and stand still.

To dwell on the moral and religious reflexions suggested by this train of thought is not to our present purpose ; but we may observe that it introduces a *homogeneity,* so to speak, into the government of the universe.   Perpetual change, perpetual progression, increase and diminution, appear to be the rules of the material world, and to prevail without exception.   The smaller portions of matter which we have near us, and the larger, which appear as luminaries at a vast distance, different as they are in our mode of conceiving them, obey the same laws of motion ; and these laws produce the same results ; in both cases motion is perpetually destroyed, except it be repaired by some living power ; in both cases the relative rest of the parts of a material system is the conclusion to which its motion tends.

4. It may perhaps appear to some, that this acknowledgment of the tendency of the system to derangement through the action of a resisting

medium is inconsistent with the argument which we have drawn in a previous chapter, from the provisions for its stability. In reality, however, the two views are in perfect agreement, so far as our purpose is concerned. The main point which we had to urge, in the consideration of the stability of the system, was, not that it is constructed to last for ever, but that while it lasts, the deviations from its mean condition are very small. It is this property which fits the world for its uses. To maintain either the past or the future eternity of the world, does not appear consistent with physical principles, as it certainly does not fall in with the convictions of the religious man, in whatever way obtained. We conceive that this state of things has had a beginning; we conceive that it will have an end. But in the mean time we find it fitted, by a number of remarkable arrangements, to be the habitation of living creatures. The conditions which secure the stability, and the smallness of the perturbations of the system, are among these provisions. If the excentricity of the orbit of Venus, or of Jupiter, were much greater than it is, not only might some of the planets, at the close of ages, fall into the sun or fly off into infinite space, but also, in the intermediate time, the earth's orbit might become much more excentric; the course of the seasons and the average of temperature might vary from what they now are, so as to injure or destroy the whole organic creation. By certain original arrangements these destructive oscillations are prevented. So long as the bodies continue to revolve, their orbits will not be much different from what they now are. And this result is not affected by the action of the resisting medium. Such a medium cannot increase the small excentricities of the orbits. The range of the periodical oscillations of heat and cold will not be extended by the mechanical effect of the medium, nor would be, even if its density were incomparably greater than it is. The resisting medium therefore does not at all counteract that which is most important in the provision for the permanency of the solar system. If the stability of the system had not been secured by the adjustments which we described in a former chapter, the course of the seasons might have been disturbed to an injurious or even destructive extent in the course of a few centuries, or even within the limits of one generation; by the effect of the resisting medium, the order of nature remains unchanged for a period, compared with which the known duration of the human race is insignificant.

But, it may be objected, the effect of the medium must be ultimately to affect the duration of the earth's revolution round the sun, and thus to derange those adaptations which depend on the length of the year. And, without question, if we permit ourselves to look forwards to that inconceivably distant period at which the effect of the medium will become sensible, this must be allowed to be true, as has been already stated. Millions, and probably millions of millions, of years express inadequately the distance of time at which this cause would produce a serious effect. That the machine of the universe is so constructed that it may answer its purposes for such a period, is surely sufficient proof of the skill of its workmanship, and of the reality of its purpose: and those persons, probably, who are best convinced that it is the work of a wise and good Creator, will be least disposed to consider the system as imperfect, because in its present condition it is not fitted for eternity.

5. The doctrine of a Resisting Medium leads us towards a point which the Nebular Hypothesis assumes ;—a *beginning* of the present order of things. There must have been a commencement of the motions now going on in the solar system. Since these motions, when once begun, would be deranged and destroyed in a period which, however large, is yet finite, it is obvious we cannot carry their origin indefinitely backwards in the range of past duration. There is a period in which these revolutions, whenever they had begun, would have brought the revolving bodies into contact with the central mass ; and this period has in our system not yet elapsed. The watch is still going, and therefore it must have been wound up within a limited time.

The solar system, at this its beginning, must

have been arranged and put in motion by some cause. If we suppose this cause to operate by means of the configurations and the properties of previously existing matter, these configurations must have resulted from some still previous cause, these properties must have produced some previous effects. We are thus led to a condition still earlier than the assumed beginning ;—to an origin of the original state of the universe ; and in this manner we are carried perpetually further and further back, through a labyrinth of mechanical causation, without any possibility of finding anything in which the mind can acquiesce or rest, till we admit " a First Cause which is not mechanical."

Thus the argument which was before urged against those in particular, who put forwards the Nebular Hypothesis in opposition to the admission of an Intelligent Creator, offers itself again, as cogent in itself, when we adopt the opinion of a resisting medium, for which the physical proofs have been found to be so strong. The argument is indeed forced upon our minds, whatever view we take of the past history of the universe. Some have endeavoured to evade its force by maintaining that the world as it now exists has existed from eternity. They assert that the present order of things, or an order of things in some way resembling the present, produced by the same causes, governed by the same laws, has prevailed through an infinite succession of past ages. We shall not dwell upon any objections to this tenet which might be drawn from our own conceptions, or from what may be called metaphysical sources. Nor shall we refer to the various considerations which history, geology, and astronomical records supply, and which tend to show, not only that the past duration of the present course of things is finite, but that it is short, compared with such periods as we have had to speak of. But we may observe, that the doctrine of a resisting medium once established, makes this imagination untenable ; compels us to go back to the origin, not only of the present course of the world, not only of the earth, but of the solar system itself; and thus sets us forth upon that path of research into the series of past causation, where we obtain no answer of which the meaning corresponds to our questions, till we rest in the conclusion of a most provident and most powerful Creating Intelligence.

It is related of Epicurus that when a boy, reading with his preceptor these verses of Hesiod,

Ητοι μεν πρωτιϛα Χαος γενετ', αυταρ επειτα
Γαι' ευρυϛερνος παντων εδος ασφαλες αιει
Αθανατων,

Eldest of beings, Chaos first arose,
Thence Earth wide stretched, the steadfast seat of all
The Immortals,

the young scholar first betrayed his inquisitive

genius by asking " And chaos whence ?" When in his riper years he had persuaded himself that this question was sufficiently answered by saying that chaos arose from the concourse of atoms, it is strange that the same inquisitive spirit did not again suggest the question " and atoms whence?" And it is clear that however often the question "whence?" had been answered, it would still start up as at first. Nor could it suffice as an answer to say, that earth, chaos, atoms, were portions of a series of changes which went back to eternity. The preceptor of Epicurus informed him, that to be satisfied on the subject of his enquiry, he must have recourse to the philosophers. If the young speculator had been told that chaos (if chaos indeed preceded the present order) was produced by an Eternal Being, in whom resided purpose and will, he would have received a suggestion which, duly matured by subsequent contemplation, might have led him to a philosophy far more satisfactory than the material scheme can ever be, to one who looks, either abroad into the universe, or within into his own bosom.

Chapter IX.

Mechanical Laws.

In the preceding observations we have supposed
the laws, by which different kinds of matter act
and are acted upon, to be already in existence ;
and have endeavoured to point out evidences of
design and adaptation, displayed in the selection
and arrangement of these materials of the universe.
These materials are, it has appeared, supplied in
such measures and disposed in such forms, that
by means of their properties and laws the business
of the world goes on harmoniously and benefi-
cially. But a further question occurs : how came
matter to have such properties and laws ? Are
these also to be considered as things of selection
and institution ? And if so, can we trace the
reasons why the laws were established in their
present form ; why the properties which matter
actually possesses were established and bestowed
upon it ? We have already attempted, in a
previous part of this work, to point out some of
the advantages which are secured by the existing laws of heat, light and moisture. Can we, in
the same manner, point out the benefits which
arise from the present constitution of those laws

of matter which are mainly concerned in the production of cosmical phenomena?

It will readily be perceived that the discussion of this point must necessarily require some effort of abstract thought. The laws and properties of which we have here to speak, the laws of motion and the universal properties of matter, are so closely interwoven with our conceptions of the external world, that we have great difficulty in conceiving them not to exist, or to exist other than they are. When we press or lift a stone, we can hardly imagine that it could, by possibility, do otherwise than resist our effort by its hardness and by its heaviness, qualities so familiar to us : when we throw it, it seems inevitable that its motion should depend on the impulse we give, just as we find that it invariably does.

Nor is it easy to say how far it is really possible to suppose the fundamental attributes of matter to be different from what they are. If we, in our thoughts, attempt to divest matter of its powers of resisting and moving, it ceases to be matter, according to our conceptions, and we can no longer reason upon it with any distinctness. And yet it is certain that we can conceive the laws of hardness and weight and motion to be quite different from what they are, and can point out some of the consequences which would result from such difference. The properties of matter, even the most fundamental and universal ones, do not obtain by any absolute necessity, resembling that which belongs to the properties of geometry. A line touching a circle is *necessarily* perpendicular to a line drawn to the centre through the point touched ; for it may be shewn that the contrary involves a contradiction. But there is no contradiction in supposing that a body's motion should naturally diminish, or that its weight should increase in removing further from the earth's centre.

Thus the properties of matter and the laws of motion are what we find them, not by virtue of any internal necessity which we can understand. The study of such laws and properties may therefore disclose to us the character of that external agency by which we conceive them to have been determined to be what they are ; and this must be the same agency by which all other parts of the constitution of the universe were appointed and ordered.

But we can hardly expect, with regard to such subjects, that we shall be able to obtain any complete or adequate view of the reasons why these general laws are so selected, and so established. These laws are the universal basis of all operations which go on, at any moment, in every part of space, with regard to every particle of matter, organic and inorganic. All other laws and properties must have a reference to these, and must be influenced by them ; both such as men have already discovered, and the far greater number which remain still unknown. The general economy and mutual relations of all parts of the universe, must be subordinate to the laws of motion and matter of which we here speak. We can easily suppose that the various processes of nature, and the dependencies of various creatures, are affected in the most comprehensive manner by these laws ;—are simplified by *their* simplicity, made consistent by *their* universality ; rendered regular by *their* symmetry. We can easily suppose that in this way there may be the most profound and admirable reasons for the existence of the present universal properties of matter, which we cannot apprehend in consequence of the limited nature of our knowledge, and of our faculties. For, compared with the whole extent of the universe, the whole aggregate of things and relations and connexions which exist in it, our knowledge is most narrow and partial, most shallow and superficial. We cannot suppose, therefore, that the reasons which we discover for the present form of the laws of nature go nearly to the full extent, or to the bottom of the reasons, which a more complete and profound insight would enable us to perceive. To do justice to such reasons, would require nothing less than a perfect acquaintance with the whole constitution of every part of creation ; a knowledge which man has not, and, so far as we can conceive, never can have.

We are certain, therefore, that our views, with regard to this part of our subject, must be imperfect and limited. Yet still man has *some* knowledge with regard to various portions of nature; and with regard to those most general and comparatively simple facts to which we now refer, his knowledge is more comprehensive, and goes deeper than it does in any other province. We conceive, therefore, that we shall not be engaged in any rash or presumptuous attempt, if we endeavour to point out some of the advantages which are secured by the present constitution of some of the general mechanical laws of nature; and to suggest the persuasion of that purpose and wise design, which the selection of such laws will thus appear to imply.

Chapter X.

The Law of Gravitation.

WE shall proceed to make a few observations on the Law of Gravity, in virtue of which the motions of planets about the sun, and of satellites about their planets take place; and by which also are produced the fall downwards of all bodies within our reach, and the pressure which they exert upon their supports when at rest. The identification of the latter forces with the former, and the discovery of the single law by which these forces are every where regulated, was the great discovery of Newton : and we wish to make it appear that this law is established by an intelligent and comprehensive selection.

The law of the sun's attraction upon the planets is, that this attraction varies *inversely* as the square of the distance ; that is, it decreases as that square increases. If we take three points or planets of the solar system, the distances of which from the sun are in proper proportion 1, 2, 3 ; the attractive force which the sun at these distances exercises, is as 1, 1-4th, and 1-9th respectively. In the smaller variations of distance which occur in the elliptical motion of one planet, the variations of the force follow the same law. Moreover, not only does the sun attract the planets, but they attract each other according to the same law ; the tendency to the earth which makes bodies heavy, is one of the effects of this law : and all these effects of the attractions of large masses may be traced to the attractions of the particles of which they are composed ; so that the final generalization, including all the derivative laws, is, that every particle of matter in the universe attracts every other, according to the law of the inverse square of the distance.

Such is the law of universal gravitation. Now, the question is, why do either the attractions of masses, or those of their component particles, follow this law of the inverse square of the distance rather than any other? When the distance becomes 1, 2, and 3, why should not the force also become 1, 2, and 3?—or if it must be weaker at points more remote from the attracting body, why should it not be 1, a half, a third? or 1, 1-8th, 1-27th? Such laws could easily be expressed mathematically, and their consequences calculated. Can any reason be assigned why the law which we find in operation *must* obtain? Can any be assigned why it *should* obtain?

The answer to this is, that no reason, at all satisfactory, can be given why such a law must, of necessity, be what it is; but that very strong reasons can be pointed out why, for the beauty and advantage of the system, the present one is better than others. We will point out some of these reasons.

1. In the first place, the system could not have subsisted, if the force had followed a *direct* instead of an inverse law, with respect to the distance; that is, if it had increased when the distance increased. It has been sometimes said, that " all direct laws of force are excluded on account of the danger from perturbing forces ;*" that if the planets had pulled at this earth, the harder the further off they were, they would have dragged it entirely out of its course. This is not an exact

* Paley.

statement of what would happen : if the force
were to be simply in the direct ratio of the distance, any number of planets might revolve in
the most regular and orderly manner. Their
mutual effects, which we may call perturbations
if we please, would be considerable ; but these
perturbations would be so combined with the
unperturbed motion, as to produce a new motion
not less regular than the other. This curious
result would follow, that every body in the system
would describe, or seem to describe, about every
other, an exact elliptical orbit ; and that the times
of the revolution of every body in its orbit would
be all equal. This is proved by Newton, in the
64th proposition of the Principia. There would
be nothing to prevent all the planets, on this
supposition, from moving round the sun in orbits
exactly circular, or nearly circular, according to
the mode in which they were set in motion.

But though the perturbations of the system
would not make this law inadmissible, there are
other circumstances which would do so. Under
this law, the gravity of bodies at the earth's surface would cease to exist. Nothing would fall or
weigh downwards. The greater action of the
distant sun and planets would exactly neutralize
the gravity of the earth : a ball thrown from the
hand, however gently, would immediately become
a satellite of the earth, and would for the future
accompany it in its course, revolving about it in
the space of one year. All terrestrial things

would float about with no principle of coherence
or stability: they would obey the general law of
the system, but would acknowledge no particular
relation to the earth. We can hardly pretend to
judge of the abstract possibility of such a system
of things; but it is clear that it could not exist
without an utter subversion of all that we can
conceive of the economy and structure of the
world which we inhabit.

With any other direct law of force, we should
in like manner lose gravity, without gaining the
theoretical regularity of the planetary motions
which we have described in the case just considered.

2. Among *inverse* laws of the distance, (that is
those according to which the force diminishes as
the distance from the origin of force increases,) all
which diminish the central force faster than the
cube of the distance increases are inadmissible,
because they are incompatible with the permanent revolution of a planet. Under such laws it
would follow, that a planet would describe a spiral
line about the sun, and would either approach
nearer and nearer to him perpetually, or perpetually go further and further off: nearly as a
stone at the end of a string, when the string is
whirled round, and is allowed to wrap round the
hand, or to unwrap from it, approaches to or
recedes from the hand.

If we endeavour to compare the law of the
inverse square of the distance, which really regu-

lates the central force, with other laws, not obviously inadmissible, as for instance, the inverse simple ratio of the distance, a considerable quantity of calculation is found to be necessary in order to trace the results, and especially the perturbations in the two cases. The perturbations in the supposed case have not been calculated; such a calculation being a process so long and laborious that it is never gone through, except for the purpose of comparing the results of theory with those of observation, as we can do with regard to the law of inverse square. We can only say, therefore, that the stability of the system, and the moderate limits of the perturbations, which we know to be secured by the existing law, would not, so far as we know, be obtained by any different law

Without going into further examination of the subject, we may observe that there are some circumstances in which the present system has a manifest superiority in its simplicity over the condition which would have belonged to it if the force had followed any other law. Thus, with the present law of gravitation the planets revolve, returning perpetually on the same track, very nearly. The earth describes an oval, in consequence of which motion she is nearer to the sun in our winter than in our summer by about one-thirtieth part of the whole distance. And, as the matter now is, the nearest approach to the sun, and the farthest recess from him, occur always at the same points of the orbit. There is indeed a slight alteration in these points arising from disturbing forces, but this is hardly sensible in the course of several ages. Now if the force had followed any other law, we should have had the earth running perpetually on a new track. The greatest and least distances would have occurred at different parts in every successive revolution. The orbit would have perpetually intersected and been interlaced with the path described in former revolutions; and the simplicity and regularity which characterises the present motion would have been quite wanting.

3. Another peculiar point of simplicity in the present law of mutual attraction is this : that it makes the law of attraction for spherical masses the same as for single particles. If particles attract with forces which are inversely as the square of the distance, spheres composed of such particles, will exert a force which follows the same law. In this character the present law is singular, among all possible laws, excepting that of the direct distance which we have already discussed. If the law of the gravitation of particles had been that of the inverse simple distance, the attraction of a sphere would have been expressed by a complex series of mathematical expressions, each representing a simple law. It is truly remarkable that the law of the inverse square of the distance, which appears to be selected as that of the

masses of the system, and of which the mechanism is, that it arises from the action of the *particles* of the system, should lead us to the same law for the action of these particles : there is a striking *prerogative* of simplicity in the law thus adopted.

The law of gravitation actually prevailing in the solar system has thus great and clear advantages over any law widely different from it ; and has moreover, in many of its consequences, a simplicity which belongs to this precise law alone. It is in many such respects a *unique* law ; and when we consider that it possesses several *properties* which are *peculiar* to it, and several *advantages* which may be peculiar to it, and which are certainly nearly so ; we have some ground, it would appear, to look upon its peculiarities and its advantages as connected. For the reasons mentioned in the last chapter, we can hardly expect to see fully the way in which the system is benefitted by the simplicity of this law, and by the mathematical elegance of its consequences : but when we see that it has some such beauties, and some manifest benefits, we may easily suppose that our ignorance and limited capacity alone prevent our seeing that there are, for the selection of this law of force, reasons of a far more refined and comprehensive kind than we can distinctly apprehend.

4. But before quitting this subject we may offer a few further observations on the question,

whether gravitation and the law of gravitation be *necessary* attributes of matter. We have spoken of the selection of this law, but is it selected? Could it have been otherwise? Is not the force of attraction a necessary consequence of the fundamental properties of matter?

This is a question which has been much agitated among the followers of Newton. Some have maintained, as Cotes, that gravity is an inherent property of all matter; others, with Newton himself, have considered it as an appendage to the essential qualities of matter, and have proposed hypotheses to account for the mode in which its effects are produced.

The result of all that can be said on the subject appears to be this : that no one can demonstrate the possibility of deducing gravity from the acknowledged fundamental properties of matter : and that no philosopher asserts, that matter has been found to exist, which was destitute of gravity. It is a property which we have no right to call *necessary* to matter, but every reason to suppose *universal*.

If we could shew gravity to be a necessary consequence of those properties which we adopt as essential to our notion of matter, (extension, solidity, mobility, inertia) we might then call it also one of the essential properties. But no one probably will assert that this is the case. Its universality is a fact of observation merely.

How then can a property,—in its existence so needful for the support of the universe, in its laws so well adapted to the purposes of creation, —how came it to be thus universal? Its being found everywhere is necessary for its uses; but this is so far from being a sufficient explanation of its existence, that it is an additional fact to be explained. We have here, then, an agency most simple in its rule, most comprehensive in its influence, most effectual and admirable in its operation. What evidence could be afforded of design, by laws of mechanical action, which this law thus existing and thus operating does not afford us?

5. It is not necessary for our purpose to consider the theories which have been proposed to account for the action of gravity. They have proceeded on the plan of reducing this action to the result of pressure or impulse. Even if such theories could be established, they could not much, or at all, affect our argument; for the arrangements by which pressure or impact could produce the effects which gravity produces, must be at least as clearly results of contrivance, as gravity itself can be.

In fact, however, none of these attempts can be considered as at all successful. That of Newton is very remarkable: it is found among the Queries in the second edition of his Optics. " To shew," he says, " that I do not take gravity for

an essential property of bodies, I have added one
question concerning its cause, choosing to pro-
pose it by way of question, because I am not yet
satisfied about it for want of experiments." The
hypothesis which he thus suggests is, that there
is an elastic medium pervading all space, and
increasing in elasticity as we proceed from dense
bodies outwards : that this " causes the gravity
of such dense bodies to each other : every body
endeavouring to go from the denser parts of the
medium towards the rarer." Of this hypothesis
we may venture to say, that it is in the first
place quite gratuitous ; we cannot trace in any
other phenomena a medium possessing these
properties : and in the next place, that the hypothesis contains several suppositions which are
more complex than the fact to be explained, and
none which are less so. Can we, on Newton's
principles, conceive an elastic medium otherwise
than as a collection of particles, repelling each
other ? and is the repulsion of such particles a
simpler fact than the attraction of those which
gravitate ? And when we suppose that the medium becomes more elastic as we proceed from
each attracting body, what cause can we conceive
capable of keeping it in such a condition, except
a repulsive force emanating from the body itself:
a supposition at least as much requiring to be
accounted for, as the attraction of the body. It
does not appear, then, that this hypothesis will

bear examination; although, for our purpose, the argument would be rather strengthened than weakened, if it could be established.

6. Another theory of the cause of gravity, which at one time excited considerable notice, was that originally proposed by M. Le Sage, in a memoir entitled " Lucrece Newtonien," and further illustrated by M. Prevost; according to which all space is occupied by currents of matter, moving perpetually in straight lines, in all directions, with a vast velocity, and penetrating all bodies. When two bodies are near each other, they intercept the current which would flow in the intermediate space if they were not there, and thus receive a tendency towards each other from the pressure of the currents on their farther sides. Without examining further this curious and ingenious hypothesis, we may make upon it the same kind of observations as before;—that it is perfectly gratuitous, except as a means of explaining the phenomena; and that, if it were proved, it would still remain to be shown what necessity has caused the existence of these *two kinds* of matter; the first kind being that which is commonly called matter, and which alone affects our senses, while it is inert as to any tendency to motion; the second kind being something imperceptible to our senses, except by the effects it produces on matter of the former kind; yet exerting an impulse on

every material body, permeating every portion of common matter, flowing with inconceivable velocity, in inexhaustible abundance, from every part of the abyss of infinity on one side, to the opposite part of the same abyss ; and so constituted that through all eternity it can never bend its path, or return, or tarry in its course.

If we were to accept this theory, it would little or nothing diminish our wonder at the structure of the universe. We might well continue to admire the evidence of contrivance, if such a machinery should be found to produce all the effects which flow from the law of gravitation.

7. The arguments for and against the necessity of the law of the inverse square of the distance in the force of gravity, were discussed with great animation about the middle of the last century. Clairault, an eminent mathematician, who did more than almost any other person for the establishment and developement of the Newtonian doctrines, maintained, at one period of his researches, not only that the inverse square was not the *necessary* law, but also that it was not the *true* law. The occasion of this controversy was somewhat curious.

Newton and other astronomers had found that the line of the moon's *apsides* (that is of her greatest and least distances from the earth) moves round to different parts of the heavens with a velocity twice as great as that which the calculation from the law of gravitation seems at first to give. According to the theory, it appeared that this line ought to move round once in eighteen years ; according to observation, it moves round once in nine years. This difference, the only obvious failure of the theory of gravitation, embarrassed mathematicians exceedingly. It is true, it was afterwards discovered that the apparent discrepancy arose from a mistake ; the calculation, which is long and laborious, was supposed to have been carried far enough to get close to the truth ; but it appeared afterwards that the residue which had been left out as insignificant, produced, by an unexpected turn in the reckoning, an effect as large as that which had been taken for the whole. But this discovery was not made till afterwards ; and in the mean time the law of the inverse square appeared to be at fault. Clairault tried to remedy the defect by supposing that the force of the earth's gravity consisted of a large force varying as the square of the distance, and a very small force varying as the fourth power (the square of the square). By such a supposition, observation and theory could be reconciled ; but on the suggestion of it, Buffon came forward with the assertion that the force *could* not vary according to any other law than the inverse square. His arguments are rather metaphysical than physical or mathematical. Gravity, he urges, is a quality, an emanation ; and all emanations are inversely as the square of the distance, as light, odours. To this Clairault replies by asking, how we know that light and odours have their intensity inversely as the square of the distance from their origin : not, he observes, by measuring the intensity, but by *supposing* these effects to be material emanations. But who, he asks, supposes gravity to be a material emanation *from* the attracting body.

Buffon again pleads that so many facts prove the law of the inverse square, that a single one, which occurs to interfere with this agreement, must be in some manner capable of being explained away. Clairault replies, that the facts do *not* prove this law to obtain exactly ; that small effects, of the same order as the one under discussion, have been neglected ; and that therefore the law is only known to be true, *as far* as such an approximation goes, and no farther.

Buffon then argues, that there can be no such additional fraction of the force, following a different law, as Clairault supposes : for what, he asks, is there to determine the magnitude of the fraction to one amount rather than another ? why should nature select for it any particular magnitude ? To this it is replied, that, whether we can explain the fact or not, nature does select certain magnitudes in preference to others : that where we ascertain she does this, we are not to deny the fact because we cannot assign the grounds of her preference. What is there, it is asked, to determine the magnitude of the whole force at any fixed distance? We cannot tell; yet the force is of a certain definite intensity and no other.

Finally Clairault observes, that we have, in cohesion, capillary attraction, and various other cases, examples of forces varying according to other laws than the inverse square; and that therefore this cannot be the only possible law.

The discrepancy between observation and theory which gave rise to this controversy was removed, as has been already stated, by a more exact calculation: and thus, as Laplace observes, in this case the metaphysician turned out to be right and the mathematician to be wrong. But most persons, probably, who are familiar with such trains of speculation, will allow, that Clairault had the best of the argument, and that the attempts to show the law of gravitation to be necessarily what it is, are fallacious and unsound.

8. We may observe, however, that the law of gravitation according to the inverse square of the distance, which thus regulates the motions of the solar system, is not confined to that province of the universe, as has been shewn by recent researches. It appears by the observations and calculations of Sir John Herschel, that several

of the stars, called *double stars*, consist of a pair of luminous bodies which revolve above each other in ellipses, in such a manner as to shew that the force, by which they are attracted to each other, varies according to the law of the inverse square. We thus learn a remarkable fact concerning bodies which seemed so far removed that no effort of our science could reach them; and we find that the same law of mutual attraction which we have before traced to the farthest bounds of the solar system, prevails also in spaces at a distance compared with which the orbit of Saturn shrinks into a point. The establishment of such a truth certainly suggests, as highly probable, the prevalence of this law among all the bodies of the universe. And we may therefore suppose, that the same ordinance which gave to the parts of our system that rule by which they fulfil the purposes of their creation, impressed the same rule on the other portions of matter which are scattered in the most remote parts of the universe; and thus gave to their movements the same grounds of simplicity and harmony which we find reason to admire, as far as we can acquire any knowledge of our own more immediate neighbourhood.

CHAPTER XI.

The Laws of Motion.

WE shall now make a few remarks on the general Laws of Motion by which all mechanical effects take place. Are we to consider these as instituted laws? and if so, can we point out any of the reasons which we may suppose to have led to the selection of those laws which really exist?

The observations formerly made concerning the inevitable narrowness and imperfection of our conclusions on such subjects, apply here, even more strongly than in the case of the law of gravitation. We can hardly conceive matter divested of these laws; and we cannot perceive or trace a millionth part of the effects which they produce. We cannot, therefore, expect to go far in pointing out the advantages of these laws such as they now obtain.

It would be easy to show that the fundamental laws of motion, in whatever form we state them, possess a very preeminent simplicity, compared with almost all others, which we might imagine as existing. This simplicity has indeed produced an effect on men's minds which, though delusive, appears to be very natural; several writers have treated these laws as self-evident, and necessarily flowing from the nature of our conceptions. We conceive that this is an erroneous view, and that these laws are known to us to be what they are, by experience only; that they might, so far as we can discern, have been any others. They appear therefore to be selected for their fitness to answer their purposes; and we may, perhaps, be able to point out some instances in which this fitness is apparent to us.

Newton, and many English philosophers, teach the existence of *three* separate fundamental laws of motion, while most of the eminent mathematicians of France reduce these to *two*, the law of inertia and the law that force is proportional to velocity. As an example of the views which we wish to illustrate, we may take the law of inertia, which is identical with Newton's first Law of Motion. This law asserts, that a body at rest continues at rest, and that a body in motion goes on moving with its velocity and direction unchanged, except so far as it is acted on by extraneous forces.*

---

* If the Laws of Motion are stated as *three*, which we conceive to be the true view of the subject, the other two, as applied in mechanical reasonings, are the following:

Second Law. When a force acts on a body in motion, it produces the same effect as if the same force acted on a body at rest.

Third Law. When a force of the nature of pressure produces motion, the velocity produced is proportional to the force, other things being equal.

We conceive that this law, simple and universal as it is, cannot be shown to be necessarily true. It might be difficult to discuss this point in general terms with any clearness ; but let us take the only example which we know of a motion absolutely uniform, in consequence of the absence of any force to accelerate or retard it ;— this motion is the rotation of the earth on its axis.

1. It is scarcely possible that discussions on such subjects should not have a repulsive and scholastic aspect, and appear like disputes about words rather than things. For mechanical writers have exercised all their ingenuity so to circumscribe their notions and so to define their terms that these fundamental truths should be expressed in the simplest manner : the consequence of which has been, that they have been made to assume the appearance rather of identical assertions than of general facts of experience. But in order to avoid this inconvenience, as far as may be, let us take the *first law of motion* as exemplified in a particular case, the rotation of the earth. Of all the motions with which we are acquainted this is alone invariable. Each day, measured by the passages of the stars, is so precisely of the same length that, according to Laplace's calculations, it is impossible that a difference of one hundredth of a second of time should have obtained between the length of the

day in the earliest ages and at the present time. Now why is this? How is this very remarkable uniformity preserved in this particular phenomenon, while all the other motions of the system are subject to inequalities? How is it that in the celestial machine no retardation takes place by the lapse of time, as would be the case in any machine which it would be possible for human powers to construct? The answer is, that in the earth's revolution on her axis no cause operates to retard the speed, like the imperfection of materials, the friction of supports, the resistance of the ambient medium; impediments which cannot, in any human mechanism, however perfect, be completely annihilated. But here we are led to ask again, why should the speed continue the same when not affected by an extraneous cause? why should it not languish and decay of itself by the mere lapse of time? That it might do so, involves no contradiction, for it was the common, though erroneous, belief of all mechanical speculators, to the time of Galileo. We can conceive velocity to diminish in proceeding from a certain point of time, as easily as we can conceive force to diminish in proceeding from a certain point of space, which in attractive forces really occurs. But, it is sometimes said, the *motion* (that is the velocity) *must* continue the same from one instant to another, for there is nothing to change it.

This appears to be taking refuge in words. We may call the velocity, that is the speed of a body, its motion ; but we cannot, by giving it this name, make it a *thing* which has any *à priori* claim to permanence, much less any self-evident constancy. Why must the speed of a body, left to itself, continue the same, any more than its temperature. Hot bodies grow cooler of themselves, why should not quick bodies go slower of themselves ? Why must a body describe 1000 feet in the next second because it has described 1000 feet in the last ? Nothing but experience, under proper circumstances, can inform us whether bodies, abstracting from external agency, do move according to such a rule. We find that they do so, we learn that all diminution of their speed which ever takes place, can be traced to external causes. Contrary to all that men had guessed, motion appears to be of itself endless and unwearied. In order to account for the unalterable permanence of the length of our day, all that is requisite is to show that there is no let or hindrance in the way of the earth's rotation ;—no resisting medium or alteration of size,—she " spinning *sleeps*" on her axle, as the poet expresses it, and may go on sleeping with the same regularity for ever, so far as the experimental properties of motion are concerned.

Such is the necessary consequence of the first

law of motion ; but the law itself has no necessary existence, so far as we can see. It was discovered only after various perplexities and false conjectures of speculators on mechanics. We have learnt that it is so, but we have not learnt, nor can any one undertake to teach us, that it must have been so. For aught we can tell, it is one among a thousand equally possible laws, which might have regulated the motions of bodies.

2. But though we have thus no reason to consider this as the only possible law, we have good reason to consider it as the best, or at least as possessing all that we can conceive of advantage. It is the *simplest* conceivable of such laws. If the velocity had been compelled to change with the time, there must have been a law of the change, and the kind and amount of this change must have been determined by its dependance on the time and other conditions. This, though quite supposable, would undoubtedly have been more complex than the present state of things. And though complexity does not appear to embarrass the operations of the laws of nature, and is admitted, without scruple, when there is reason for it, simplicity is the usual character of such laws, and appears to have been a ground of selection in the formation of the universe, as it is a mark of beauty to us in our contemplation of it.

But there is a still stronger apparent reason for the selection of this law of the preservation of motion. If the case had been otherwise, the universe must necessarily in the course of ages have been reduced to a state of rest, or at least to a state not sensibly differing from it. If the earth's motion, round its axis, had slackened by a very small quantity, for instance, by a hundredth of a second in a revolution, and in this proportion continued, the day would have been already lengthened by six hours in the 6000 years which have elapsed since the history of the world began; and if we suppose a longer period to precede or to follow, the day might be increased to a month or to any length. All the adaptations which depend on the length of the day would consequently be deranged. But this would not be all; for the same law of motion is equally requisite for the preservation of the annual motion of the earth. If her motion were retarded by the establishment of any other law instead of the existing one, she would wheel nearer and nearer to the sun at every revolution, and at last reach the centre, like a falling hoop. The same would happen to the other planets; and the whole solar system would, in the course of a certain period, be gathered into a heap of matter without life or motion. In the present state of things on the other hand, the system, as we have already explained, is, by a combination of remarkable provisions, calculated for an almost indefinite existence, of undiminished fitness for its purposes.

There are, therefore, manifest reasons, why, of all laws which could occupy the place of the first law of motion, the one which now obtains is the only one consistent with the durability and uniformity of the system ;—the one, therefore, which we may naturally conceive to be selected by a wise contriver. And as, along with this, it has appeared that we have no sort of right to attribute the establishment of this law to anything but selection, we have here a striking evidence, to lead us to a perception of that Divine mind, by which means so simple are made to answer purposes so extensive and so beneficial.

CHAPTER XII.

*Friction.**

WE shall not pursue this argument of the last chapter, by considering the other laws of motion in the same manner as we have there considered the first, which might be done. But the facts

---

* Though Friction is not concerned in any cosmical phenomena, we have thought this the proper place to introduce the consideration of it; since the contrast between the cases in which it does act, and those in which it does not, is best illustrated by a comparison of cosmical with terrestrial motions.

which form exceptions and apparent contradictions to the first law of which we have been treating, and which are very numerous, offer, we conceive, an additional exemplification of the same argument; and this we shall endeavour to illustrate.

The rule that a body naturally moves for ever with an undiminished speed, is so far from being obviously true, that it appears on a first examination to be manifestly false. The hoop of the school boy, left to itself, runs on a short distance, and then stops; his top spins a little while, but finally flags and falls; all motion on the earth appears to decay by its own nature; all matter which we move appears to have a perpetual tendency to divest itself of the velocity which we communicate to it. How is this reconcileable with the first law of motion on which we have been insisting?

It is reconciled principally by considering the effect of *Friction*. Among terrestrial objects friction exerts an agency almost as universal and constant as the laws of motion themselves; an agency which completely changes and disguises the results of those laws. We shall consider some of these effects.

It is probably not necessary to explain at any length the nature and operation of friction. When a body cannot move without causing two surfaces to rub together, this rubbing has a tendency to diminish the body's motion or to prevent it entirely. If the body of a carriage be placed on the earth without the wheels, a considerable force will be requisite in order to move it at all: it is here the friction against the ground which obstructs the motion. If the carriage be placed on its wheels, a much less force will move it, but if moved it will soon stop: it is the friction at the ground and at the axles which stops it: placed on a level rail road, with well made and well oiled wheels, and once put in motion, it might run a considerable distance alone, for the friction is here much less; but there *is* friction, and therefore the motion would after a time cease.

1. The friction which we shall principally consider is the friction which *prevents* motion. So employed, friction is one of the most universal and important agents in the mechanism of our daily comforts and occupations. It is a force which is called into play to an extent incomparably greater than all the other forces with which we are concerned in the course of our daily life. We are dependent upon it at every instant and in every action: and it is not possible to enumerate the ways in which it serves us; scarcely even to suggest a sufficient number of them to give us a true notion of its functions.

What can appear a more simple operation than standing and walking? yet it is easy to see that without the aid of friction these simple actions would scarcely be possible. Every one knows how difficult and dangerous they are when performed on smooth ice. In such a situation we cannot always succeed in standing : if the ice be very smooth, it is by no means easy to walk, even when the surface is perfectly level ; and if it were ever so little inclined, no one would make the attempt. Yet walking on the ice and on the ground differ only in our experiencing more friction in the latter case. We say *more*, for there is a considerable friction even in the case of ice, as we see by the small distance which a stone slides when thrown along the surface. It is this friction of the earth which, at every step we take, prevents the foot from sliding back ; and thus allows us to push the body and the other foot forwards. And when we come to violent bodily motions, to running, leaping, pulling or pushing objects, it is easily seen how entirely we depend upon the friction of the ground for our strength and force. Every one knows how completely powerless we become in any of these actions by the *foot slipping*.

In the same manner it is the friction of objects to which the hand is applied, which enables us to hold them with any degree of firmness. In some contests it was formerly the custom for the combatants to rub their bodies with oil, that the adversary might not be able to keep his grasp.

If the pole of the boatman, the rope of the sailor, were thus smooth and lubricated, how weak would be the thrust and the pull! Yet this would only be the removal of friction.

Our buildings are no less dependent on this force for their stability. Some edifices are erected without the aid of cement; and if the stones be large and well squared, such structures may be highly substantial and durable; even when rude and slight, houses so built answer the purposes of life. These are entirely upheld by friction, and without that agent they would be thrown down by the Zephyr, far more easily than if all the stones were lumps of ice with a thawing surface. But even in cases where cement *binds* the masonry, it does not take the duty of *holding* it together. In consequence of the existence of friction, there is no constant tendency of the stones to separate; they are in a state of repose. If this were not so, if every shock and every breeze required to be counteracted by the cement, no composition exists which would long sustain such a wear and tear. The cement excludes the corroding elements, and helps to resist extraordinary violence; but it is friction which gives the habitual state of rest.

We are not to consider friction as a *small* force, slightly modifying the effects of other agencies. On the contrary its amount is in most cases very great. When a body lies loose on the ground, the friction is equal to one third or one half, or in some cases the whole of its weight. But in cases of bodies supported by oblique pressure, the amount is far more enormous. In the arch of a bridge, the friction which is called into play between two of the vaulting stones, may be equal to the whole weight of the bridge. In such cases this conservative force is so great, that the common theory, which neglects it, does not help us even to guess what will take place. According to the theory, certain forms of arches only will stand, but in practice almost any form will stand, and it is not easy to construct a model of a bridge which will fall.

We may see the great force of friction in the *brake*, by which a large weight running down a long inclined plane has its motion moderated and stopt; in the windlass, where a few coils of the rope round a cylinder sustain the stress and weight of a large iron anchor; in the nail or screw which holds together large beams; in the mode of raising large blocks of granite by an iron rod driven into a hole in the stone. Probably no greater forces are exercised in any processes in the arts than the force of friction; and it is always employed to produce rest, stability, moderate motion. Being always ready and never wearied, always at hand and augmenting with the exigency, it regulates, controls, subdues all motions; — counteracts all other agents;—and finally gains the mastery over all other terrestrial agencies, however violent, frequent, or long continued. The perpetual action of all other terrestrial forces appears, on a large scale, only as so many interruptions of the constant and stationary rule of friction.

The objects which every where surround us, the books or dishes which stand on our tables; our tables and chairs themselves, the loose clods and stones in the field, the heaviest masses produced by nature or art, would be in a perpetual motion, quick or slow according to the forces which acted on them, and to their size, if it were not for the tranquillizing and steadying effects of the agent we are considering. Without this, our apartments, if they kept their shape, would exhibit to us articles of furniture, and of all other kinds, sliding and creeping from side to side with every push and every wind, like loose objects in a ship's cabin, when she is changing her course in a gale.

Here, then, we have a force, most extensive and incessant in its operation, which is absolutely essential to the business of this terrestrial world, according to any notion which we can form. The more any one considers its effects, and the more he will find how universally dependent he is upon it, in every action of his life; resting or moving, dealing with objects of art or of nature, with instruments of enjoyment or of action.

2. Now we have to observe concerning this agent, Friction, that we have no ground for asserting it to be a necessary result of other properties of matter, for instance, of their solidity and coherency. Philosophers have not been able to deduce the laws of friction from the other known properties of matter, nor even to explain what we know experimentally of such laws, (which is not much,) without introducing new hypotheses concerning the surfaces of bodies, &c.—hypotheses which are not supplied us by any other set of phenomena. So far as our knowledge goes, friction is a separate property, and may be conceived to have been bestowed upon matter for particular purposes. How well it answers the purpose of fitting matter for the uses of the daily life of man, we have already seen.

We may make suppositions as to the mode in which friction is connected with the texture of bodies; but little can be gained for philosophy, or for speculation of any kind, by such conjectures respecting unknown connexions. If, on the other hand, we consider this property of friction, and find that it prevails there, and there only, where the general functions, analogies, and relations of the universe require it, we shall probably receive a strong impression that it was introduced into the system of the world *for a purpose.*

3. It is very remarkable that this force, which is thus so efficacious and discharges such im-

portant offices in all earthly mechanism, disappears altogether when we turn to the mechanism
of the heavens. All motions on the earth soon
stop ;—a machine which imitates the movements
of the stars cannot go long without winding up : but
the stars themselves have gone on in their courses
for ages, with no diminution of their motions, and
offer no obvious prospect of any change. This
is so palpable a fact, that the first attempts of
men to systematize their mechanical notions were
founded upon it. The ancients held that motions
were to be distinguished into *natural* motions and
violent,—the former go on without diminution—
the latter are soon extinguished ;—the motions of
the stars are of the former kind ;—those of a
stone thrown, and in short all terrestrial motions,
of the latter. Modern philosophers maintain that
the laws of motion are the same for celestial and
terrestrial bodies ;—that all motions are *natural*
according to the above description ;—but that in
terrestrial motions, friction comes in and alters
their character,—destroys them so speedily that
they appear to have existed only during an effort.
And that this is the case will not now be contested. Is it not then somewhat remarkable that
the same laws which produce a state of permanent
motion in the heavens, should, on the earth, give
rise to a condition in which rest is the rule and
motion the exception ? The air, the waters, and
the lighter portions of matter are, no doubt, in a

state of perpetual motion ; over these friction
has no empire : yet even their motions are interrupted, alternate, variable, and on the whole
slight deviations from the condition of equilibrium.
But in the solid parts of the globe, rest pre-
dominates incomparably over motion : and this,
not only with regard to the portions which cohere
as parts of the same solid ; for the whole surface
of the earth is covered with loose masses, which,
if the power of friction were abolished, would
rush from their places and begin one universal
and interminable dance, which would make the
earth absolutely uninhabitable.

If, on the other hand, the dominion of friction
were extended in any considerable degree into
the planetary spaces, there would soon be an end
of the system. If the planet had moved in a
fluid, as the Cartesians supposed, and if this
fluid had been subject to the rules of friction
which prevail in terrestrial fluids, their motions
could not have been of long duration. The solar
system must soon have ceased to be a system of
revolving bodies.

But friction is neither abolished on the earth,
nor active in the heavens. It operates where it is
wanted, it is absent where it would be prejudicial.
And both these circumstances occasion, in a remarkable manner, the steadiness of the course of
nature. The stable condition of the objects in
man's immediate neighbourhood, and the un-

varying motions of the luminaries of heaven, are alike conducive to his well-being. This requires that he should be able to depend upon a fixed order of place, a fixed course of time. It requires, therefore, that terrestrial objects should be affected by friction, and that celestial should not; as is the case, in fact. What further evidence of benevolent design could this part of the constitution of the universe supply?

4. There is another view which may be taken of the forces which operate on the earth to produce permanency or change. Some parts of the terrestrial system are under the dominion of powers which act energetically to prevent all motion, as the crystalline forces by which the parts of rocks are bound together; other parts are influenced by powers which produce a perpetual movement and change in the matter of which they consist; thus plants and animals are in a constant state of internal movement, by the agency of the vital forces. In the former case rigid immutability, in the latter perpetual developement, are the tendencies of the agencies employed. Now in the case of objects affected by friction, we have a kind of intermediate condition, between the constantly fixed and the constantly moveable. Such objects can and do move; but they move but for a short time if left to the laws of nature. When at rest, they can easily be put in motion, but still not with un-

limited ease ; a certain finite effort, different in different cases, is requisite for their purpose. Now this immediate condition, this capacity of receiving readily and alternately the states of rest and motion, is absolutely requisite for the nature of man, for the exertion of will, of contrivance, of foresight, as well as for the comfort of life and the conditions of our material existence. If all objects were fixed and immoveable, as if frozen into one mass ; or if they were susceptible of such motions only as are found in the parts of vegetables, we attempt in vain to conceive what would come of the business of the world. But besides the state of a particle which cannot be moved, and of a particle which cannot be stopt, we have the state of a particle moveable but not moved ; or moved, but moved only while we choose : and this state is that about which the powers, the thoughts, and the wants of man are mainly conversant.

Thus the forces by which solidity and by which organic action are produced, the laws of permanence and of developement, do not bring about all that happens. Besides these, there is a mechanical condition, that of a body exposed to friction, which is neither one of absolute permanency nor one naturally progressive ; but is yet one absolutely necessary to make material objects capable of being instruments and aids to man ; and this is the condition of by far the

greater part of terrestrial things. The habitual
course of events with regard to motion and rest
is not the same for familiar moveable articles, as
it is for the parts of the mineral, or of the
vegetable world, when left to themselves; such
articles are in a condition far better adapted than
any of those other conditions would be, to their
place and purpose. Surely this shows us an
adaptation, an adjustment, of the constitution of
the material world to the nature of man. And
as the organization of plants cannot be conceived
otherwise than as having their life and growth
for its object, so we cannot conceive that friction
should be one of the leading agencies in the
world in which man is placed, without supposing
that it was intended to be of use when man
should walk and run, and build houses and
ships, and bridges, and execute innumerable
other processes, all of which would be impossible,
admirably constituted as man is in other respects,
if friction did not exist. And believing, as we
conceive we cannot but believe, that the laws of
motion and rest were thus given with reference to
their ends, we perceive in this instance, as in
others, how wide and profound this reference is,
how simple in its means, how fertile in its consequences, how effective in its details.

BOOK III.

RELIGIOUS VIEWS.

THE contemplation of the material universe exhibits God to us as the author of the laws of material nature; bringing before us a wonderful spectacle, in the simplicity, the comprehensiveness, the mutual adaptation of these laws, and in the vast variety of harmonious and beneficial effects produced by their mutual bearing and combined operation. But it is the consideration of the moral world, of the results of our powers of thought and action, which leads us to regard the Deity in that light in which our relation to him becomes a matter of the highest interest and importance. We perceive that man is capable of referring his actions to principles of right and wrong; that both his faculties and his virtues may be unfolded and advanced by the discipline which arises from the circumstances of human society; that good men can be discriminated from the bad, only by a course of trial, by struggles with difficulty and temptation; that the best men feel deeply the need of relying, in such conflicts, on the thought of a superintending Spiritual Power; that our views of justice, our capacity for intellectual and moral advancement, and a crowd of hopes and anticipations which rise in our bosoms unsought, and cling there with inexhaustible tenacity, will not allow us to acquiesce in the belief that this life is the end of our existence. We are thus led to see that our relation to the Superintender of our moral being, to the Depositary of the supreme law of just and right, is a relation of incalculable consequence. We find that we cannot be permitted to be merely contemplators and speculators with regard to the Governor of the moral world; we must obey His will; we must turn our affections to Him; we must advance in His favour; or we offend against the nature of our position in the scheme of which He is the author and sustainer.

It is far from our purpose to represent natural religion, as of itself sufficient for our support and guidance; or to underrate the manner in which our views of the Lord of the universe have been, much more, perhaps, than we are sometimes aware, illustrated and confirmed by lights derived from revelation. We do not here speak of the manner in which men have come to believe in God, as the Governor of the moral world; but of the fact, that by the aid of one or both of these two guides, Reason or Revelation, reflecting persons in every age have been led to such a belief. And we conceive it may be useful to point out some connexion between such a belief of a just and holy Governor, and the conviction, which we have already endeavoured to impress upon the reader, of a wise and benevolent Creator of the physical world. This we shall endeavour to do in the present book.

At the same time that men have thus learnt to look upon God as their Governor and Judge, the source of their support and reward, they have also been led, not only to ascribe to him power and skill, knowledge and goodness, but also to attribute to him these qualities in a mode and degree excluding all limit:—to consider him as almighty, allwise, of infinite knowledge and inexhaustible goodness; every where present and active, but incomprehensible by our minds, both in the manner of his agency, and the degree of his perfections. And this impression concerning the Deity appears to be that which the mind receives from all objects of contemplation and all modes of advance towards truth. To this conception it leaps with alacrity and joy, and in this it acquiesces with tranquil satisfaction and growing confidence; while any other view of the nature of the Divine Power which formed and sustained the world, is incoherent and untenable, exposed to insurmountable objections and intolerable incongruities. We shall endeavour to show that the modes of employment of the thoughts to which the well conducted study of nature gives rise, do tend, in all their forms, to produce or strengthen this impression on the mind ; and that such an impression, and no other, is consistent with the widest views and most comprehensive aspects of nature and of philosophy, which our Natural Philosophy opens to us. This will be the purpose of the latter part of the present book. In the first place we shall proceed with the object first mentioned, the connexion which may be perceived between the evidences of creative power, and of moral government, in the world.

CHAPTER I.

The Creator of the Physical World is the Governor of the Moral World.

WITH our views of the moral government of the world and the religious interests of man, the study of material nature is not and cannot be directly and closely connected. But it may be of some service to trace in these two lines of reasoning, seemingly so remote, a manifest convergence to the same point, a demonstrable unity of result. It may be useful to show that we are thus led, not to two rulers of the universe, but to one God ;—to make it appear that the Creator and Preserver of the world is also the Governor and Judge of men ;—that the Author of the Laws of Nature is also the Author of the

Law of Duty ;—that He who regulates corporeal things by properties of attraction and affinity and assimilating power, is the same Being who regulates the actions and conditions of men, by the influence of the feeling of responsibility, the perception of right and wrong, the hope of happiness, the love of good.

The conviction that the Divine attributes which we are taught by the study of the material world, and those which we learn from the contemplation of man as a responsible agent, belong to the same Divine Being, will be forced upon us, if we consider the manner in which all the parts of the universe, the corporeal and intellectual, the animal and moral, are connected with each other. In each of these provinces of creation we trace refined adaptations and arrangements which lead us to the Creator and Director of so skilful a system ; but these provinces are so intermixed, these different trains of contrivance so interwoven, that we cannot, in our thoughts, separate the author of one part from the author of another. The Creator of the Heavens and of the Earth, of the inorganic and of the organic world, of animals and of man, of the affections and the conscience, appears inevitably to be one and the same God.

We will pursue this reflexion a little more into detail.

1. The *Atmosphere* is a mere mass of fluid

floating on the surface of the ball of the earth;
it is one of the inert and inorganic portions of
the universe, and must be conceived to have been
formed by the same Power which formed the solid
mass of the earth and all other parts of the solar
system.   But how far is the atmosphere from
being inert in its effects on organic beings, and
unconnected with the world of life!   By what
wonderful adaptations of its mechanical and
chemical properties, and of the vital powers of
plants, to each other, are the developement and
well-being of plants and animals secured!   The
creator of the atmosphere must have been also the
creator of plants and animals : we cannot for an
instant believe the contrary.   But the atmosphere
is not only subservient to the life of animals, and
of man among the rest; it is also the vehicle of
voice ; it answers the purpose of intercourse ;
and, in the case of man, of rational intercourse.
We have seen how remarkably the air is fitted
for this office ; the construction of the organs of
articulation, by which they are enabled to por-
form their part of the work, is, as is well known,
a most exquisite system of contrivances.   But
though living in an atmosphere capable of trans-
mitting articulate sound, and though provided
with organs fitted to articulate, man would never
attain to the use of language, if he were not
also endowed with another set of faculties.   The
powers of abstraction and generalisation, memory

and reason, the tendencies which occasion the inflexions and combinations of words, are all necessary to the formation and use of language. Are not these parts of the same scheme of which the bodily faculties by which we are able to speak are another part? Has man his mental powers independently of the creator of his bodily frame? To what purpose then, or by what cause was the curious and complex machinery of the tongue, the glottis, the larynx produced? These are useful for speech, and full of contrivances which suggest such a use as the end for which those organs were constructed. But speech appears to have been no less contemplated in the intellectual structure of man. The processes of which we have spoken, generalization, abstraction, reasoning, have a close dependence on the use of speech. These faculties are presupposed in the formation of language, but they are developed and perfected by the use of language. The mind of man then, with all its intellectual endowments, is the work of the same artist by whose hands his bodily frame was fashioned; as his bodily faculties again are evidently constructed by the maker of those elements on which their action depends. The creator of the atmosphere and of the material universe is the creator of the human mind, and the author of those wonderful powers of thinking, judging, inferring, discovering, by which we are able to reason concerning the world in which we are placed ; and which aid us in lifting our thoughts to the source of our being himself.

2. *Light*, or the means by which light is propagated, is another of the inorganic elements which forms a portion of the mere material world. The luminiferous ether, if we adopt that theory, or the fluid light of the theory of emission, must indubitably pervade the remotest regions of the universe, and must be supposed to exist, as soon as we suppose the material parts of the universe to be in existence. The origin of light then must be at least as far removed from us as the origin of the solar system. Yet how closely connected are the properties of light with the structure of our own bodies ! The mechanism of the organs of vision and the mechanism of light are, as we have seen, most curiously adapted to each other. We must suppose, then, that the same power and skill produced one and the other of these two sets of contrivances, which so remarkably *fit into* each other. The creator of light is the author of our visual powers. But how small a portion does mere visual perception constitute of the advantages which we derive from vision ! We possess ulterior faculties and capacities by which sight becomes a source of happiness and good to man. The sense of beauty, the love of art, the pleasure arising from the contemplation of nature, are all dependent on

the eye; and we can hardly doubt that these faculties were bestowed on man to further the best interests of his being. The sense of beauty both animates and refines his domestic tendencies; the love of art is a powerful instrument for raising him above the mere cravings and satisfactions of his animal nature; the expansion of mind which rises in us at the sight of the starry sky, the cloud-capt mountain, the boundless ocean, seems intended to direct our thoughts by an impressive though indefinite feeling, to the Infinite Author of All. But if these faculties be thus part of the scheme of man's inner being, given him by a good and wise creator, can we suppose that this creator was any other than the creator also of those visual organs, without which the faculties could have no operation and no existence? As clearly as light and the eye are the work of the same author, so clearly also do our capacities for the most exalted visual pleasures, and the feelings flowing from them, proceed from the same Divine Hand.

3. The creator of the earth must be conceived to be the author also of all those qualities in the soil, chemical and whatever else, by which it supports vegetable life, under all the modifications of natural and artificial condition. Among the attributes which the earth thus possesses, there are some which seem to have an especial reference to man in a state of society. Such are

the power of the earth to increase its produce
under the influence of cultivation, and the necessary existence of property in land, in order
that this cultivation may be advantageously
applied ; the rise, under such circumstances, of
a *surplus* produce, of a quantity of subsistence
exceeding the wants of the cultivators alone ;
and the consequent possibility of inequalities of
rank, and of all the arrangements of civil society.
These are all parts of the constitution of the
earth. But these would all remain mere idle
possibilities, if the nature of man had not a corresponding direction. If man had not a social
and economical tendency, a disposition to con-
gregate and cooperate, to distribute possessions
and offices among the members of the community,
to make and obey and enforce laws, the earth
would in vain be ready to respond to the care of
the husbandman. Must we not then suppose
that this attribute of the earth was bestowed upon
it by Him who gave to man those corresponding
attributes, through which the apparent niggardli-
ness of the soil is the source of general comfort
and security, of polity and law ? Must we not
suppose that He who created the soil also in-
spired man with those social desires and feelings
which produce cities and states, laws and institutions, arts and civilization ; and that thus the
apparently inert mass of earth is a part of the
same scheme as those faculties and powers with

which man's moral and intellectual progress is most connected?

4. Again :—It will hardly be questioned that the author of the material elements is also the author of the structure of animals, which is adapted to and provided for by the constitution of the elements in such innumerable ways. But the author of the bodily structure of animals must also be the author of their instincts, for without these the structure would not answer its purpose. And these instincts frequently assume the character of affections in a most remarkable manner. The love of offspring, of home, of companions, are often displayed by animals, in a way that strikes the most indifferent observer ; and yet these affections will hardly be denied to be a part of the same scheme as the instincts by which the same animals seek food and the gratifications of sense. Who can doubt that the anxious and devoted affection of the mother-bird for her young after they are hatched, is a part of the same system of Providence as the instinct by which she is impelled to sit upon her eggs ? and this, of the same by which her eggs are so organized that incubation leads to the birth of the young animal ? Nor, again, can we imagine that while the structure and affections of animals belong to one system of things, the affections of man, in many respects so similar to those of animals, and connected with the bodily frame in a manner so

closely analogous, can belong to a different
scheme. Who, that reads the touching instances
of maternal affection, related so often of the
women of all nations, and of the females of all
animals, can doubt that the principle of action is
the same in the two cases, though enlightened in
one of them by the rational faculty? And who
can place in separate provinces the supporting
and protecting love of the father and of the
mother? or consider as entirely distinct from
these, and belonging to another part of our
nature, the other kinds of family affection? or
disjoin man's love of his home, his clan, his tribe,
his country, from the affection which he bears to
his family? The love of offspring, home, friends,
in man, is then part of the same system of contrivances of which bodily organization is another
part. And thus the author of our corporeal frame
is also the author of our capacity of kindness and
resentment, of our love and of our wish to be
loved, of all the emotions which bind us to individuals, to our families, and to our kind.

It is not necessary here to follow out and
classify these emotions and affections; or to
examine how they are combined and connected
with our other motives of action, mutually giving
and receiving strength and direction. The desire
of esteem, of power, of knowledge, of society, the
love of kindred, of friends, of our country, are
manifestly among the main forces by which man

is urged to act and to abstain. And as these parts of the constitution of man are clearly intended, as we conceive, to impel him in his appointed path ; so we conceive that they are no less clearly the work of the same great Artificer who created the heart, the eye, the hand, the tongue, and that elemental world in which, by means of these instruments, man pursues the objects of his appetites, desires, and affections.

5. But if the Creator of the world be also the author of our intellectual powers, of our feeling for the beautiful and the sublime, of our social tendencies, and of our natural desires and affections, we shall find it impossible not to ascribe also to Him the higher directive attributes of our nature, the conscience and the religious feeling, the reference of our actions to the rule of duty and to the will of God.

It would not suit the plan of the present treatise to enter into any detailed analysis of the connexion of these various portions of our moral constitution. But we may observe that the existence and universality of the conception of duty and right cannot be doubted, however men may differ as to its original or derivative nature. All men are perpetually led to form judgments concerning actions, and emotions which lead to action, as right or wrong ; as what they *ought* or *ought not* to do or feel. There is a faculty which approves and disapproves, acquits or condemns the workings of our other faculties. Now, what shall we say of such a judiciary principle, thus introduced among our motives to action? Shall we conceive that while the other springs of action are balanced against each other by our Creator, this, the most pervading and universal regulator, was no part of the original scheme? That—while the love of animal pleasures, of power, of fame, the regard for friends, the pleasure of bestowing pleasure, were infused into man as influences by which his course of life was to be carried on, and his capacities and powers developed and exercised;—this reverence for a moral law, this acknowledgment of the obligation of duty,—a feeling which is everywhere found, and which may become a powerful, a predominating motive of action,—was given for no purpose, and belongs not to the design? Such an opinion would be much as if we should acknowledge the skill and contrivance manifested in the other parts of a ship, but should refuse to recognize the rudder as exhibiting any evidence of a purpose. Without the reverence which the opinion of right inspires, and the scourge of general disapprobation inflicted on that which is accounted wicked, society could scarcely go on; and certainly the feelings and thoughts and characters of men could not be what they are. Those impulses of nature which involve no acknowledgment of responsibility, and the play and struggle of interfering wishes, might preserve the species in some shape of existence, as we see in the case of brutes. But a person must be strangely constituted, who, living amid the respect for law, the admiration for what is good, the order and virtues and graces of civilized nations, (all which have their origin in some degree in the feeling of responsibility) can maintain that all these are casual and extraneous circumstances, no way contemplated in the formation of man ; and that a condition in which there should be obligation in law, no merit in self-restraint, no beauty in virtue, is equally suited to the powers and the nature of man, and was equally contemplated when those powers were given him.

If this supposition be too extravagant to be admitted, as it appears to be, it remains then that man, intended, as we have already seen from his structure and properties, to be a discoursing, social being, acting under the influence of affections, desires, and purposes, was also intended to act under the influence of a sense of duty ; and that the acknowledgment of the obligation of a moral law is as much part of his nature, as hunger or thirst, maternal love or the desire of power ; that, therefore, in conceiving man as the work of a Creator, we must imagine his powers and character given him with an intention on the Creator's part that this sense of duty should occupy its place in his constitution as an active and thinking being : and that this directive and judiciary principle is a part of the work of the same Author who made the elements to minister to the material functions, and the arrangements of the world to occupy the individual and social affections of his living creatures.

This principle of conscience, it may further be observed, does not stand upon the same level as the other impulses of our constitution by which we are prompted or restrained. By its very nature and essence, it possesses a supremacy over all others. " Your obligation to obey this law is its being the law of your nature. That your conscience approves of and attests such a course of action is itself alone an obligation. Conscience does not only offer itself to show us the way we should walk in, but it likewise carries its own authority with it, that it is our natural guide : the guide assigned us by the author of our nature." * That we ought to do an action, is of itself a sufficient and ultimate answer to the questions, *why* we should do it ?—how we are *obliged* to do it ? The conviction of duty implies the soundest reason, the strongest obligation, of which our nature is susceptible.

We appear then to be using only language which is well capable of being justified, when we speak of this irresistible esteem for what is right, this conviction of a rule of action extending beyond the gratification of our irreflective impulses, as an impress stamped upon the human mind by the Deity himself; a trace of His nature; an indication of His will; an announcement of His purpose; a promise of His favour: and though this faculty may need to be confirmed and unfolded, instructed and assisted by other aids, it still seems to contain in itself a sufficient intimation that the highest objects of man's existence are to be attained, by means of a direct and intimate reference of his thoughts and actions to the Divine Author of his being.

Such then is the Deity to which the researches of Natural Theology point; and so far is the train of reflections in which we have engaged, from being merely speculative and barren. With the material world we cannot stop. If a superior Intelligence *have* ordered and adjusted the succession of seasons and the structure of the plants of the field, we must allow far more than this at first sight would seem to imply. We must admit still greater powers, still higher wisdom for the creation of the beasts of the forest with their faculties; and higher wisdom still and more transcendent attributes, for the creation of man. And when we reach this point, we find that it is not knowledge only, not power only, not foresight and beneficence alone, which we must attribute to the Maker of the World; but that we must consider him as the Author, in us, of a reverence for moral purity and rectitude ; and, if the author of such emotions in us, how can we conceive of Him otherwise, than that these qualities are parts of his nature ; and that he is not only wise and great, and good, incomparably beyond our highest conceptions, but also conformed in his purposes to the rule which he thus impresses upon us, that is, Holy in the highest degree which we can image to ourselves as possible.

Chapter II.

On the Vastness of the Universe.

1. The aspect of the world, even without any of the peculiar lights which science throws upon it, is fitted to give us an idea of the greatness of the power by which it is directed and governed, far exceeding any notions of power and greatness which are suggested by any other contemplation. The number of human beings who surround us— the various conditions requisite for their life, nutrition, well-being, all fulfilled ;—the way in which these conditions are modified, as we pass in thought to other countries, by climate, temperament, habit ;—the vast amount of the human population of the globe thus made up ;—yet man himself but one among almost endless tribes of

animals;—the forest, the field, the desert, the air, the ocean, all teeming with creatures whose bodily wants are as carefully provided for as his;—the sun, the clouds, the winds, all attending, as it were, on these organized beings;—a host of beneficent energies, unwearied by time and succession, pervading every corner of the earth;—this spectacle cannot but give the contemplator a lofty and magnificent conception of the Author of so vast a work, of the Ruler of so wide and rich an empire, of the Provider for so many and varied wants, the Director and Adjuster of such complex and jarring interests.

But when we take a more exact view of this spectacle, and aid our vision by the discoveries which have been made of the structure and extent of the universe, the impression is incalculably increased.

The number and variety of animals, the exquisite skill displayed in their structure, the comprehensive and profound relations by which they are connected, far exceed anything which we could in any degree have imagined. But the view of the universe expands also on another side. The earth, the globular body thus covered with life, is not the only globe in the universe. There are, circling about our own sun, six others, so far as we can judge, perfectly analogous in their nature: besides our moon and other bodies analogous to it. No one can resist the temptation to conjecture, that these globes, some of them much larger than our own, are not dead and barren;—that they are, like ours, occupied with organization, life, intelligence. To conjecture is all that we can do, yet even by the perception of such a possibility, our view of the kingdom of nature is enlarged and elevated. The outermost of the planetary globes of which we have spoken is so far from the sun, that the central luminary must appear to the inhabitants of that planet, if any there are, no larger than Venus does to us; and the length of their year will be 82 of ours.

But astronomy carries us still onwards. It teaches us that, with the exception of the planets already mentioned, the stars which we see have no immediate relation to our system. The obvious supposition is that they are of the nature and order of our sun: the minuteness of their apparent magnitude agrees, on this supposition, with the enormous and almost inconceivable distance which, from all the measurements of astronomers, we are led to attribute to them. If then these are suns, they may, like our sun, have planets revolving round them; and these may, like our planet, be the seats of vegetable and animal and rational life:—we may thus have in the universe worlds, no one knows how many, no one can guess how varied:—but however many, however varied, they are still but so many provinces in the same empire, subject to common rules, governed by a common power.

But the stars which we see with the naked eye are but a very small portion of those which the telescope unveils to us. The most imperfect telescope will discover some that are invisible without it ; the very best instrument perhaps does not show us the most remote. The number which crowd some parts of the heavens is truly marvellous. Dr. Herschel calculated that a portion of the milky way, about 10 degrees long and 2½ broad, contained 258,000. In a sky so occupied the moon would eclipse 2000 of such stars at once.

We learn too from the telescope that even in this province the variety of nature is not exhausted. Not only do the stars differ in colour and appearance, but some of them grow periodically fainter and brighter, as if they were dark on one side, and revolved on their axes. In other cases two stars appear close to each other, and in some of these cases it has been clearly established, that the two have a motion of revolution about each other ; thus exhibiting an arrangement before unguessed, and giving rise, possibly, to new conditions of worlds. In other instances again, the telescope shows, not luminous points, but extended masses of dilute light, like bright clouds, hence called *nebulæ*. Some have supposed (as we have noticed in the last book) that such nebulæ by further condensation might become suns; but for such opinions we have nothing but conjecture. Some stars again have undergone permanent changes, or have absolutely disappeared, as the celebrated star of 1572, in the constellation Cassiopea.

If we take the whole range of created objects in our own system, from the sun down to the smallest animalcule, and suppose such a system, or something in some way analogous to it, to be repeated for each of the millions of stars thus revealed to us, we have a representation of the material part of the universe, according to a view which many minds receive as a probable one; and referring this aggregate of systems to the Author of the universe, as in our own system we have found ourselves led to do, we have thus an estimate of the extent to which his creative energy would thus appear to have been exercised in the material world.

If we consider further the endless and admirable contrivances and adaptations which philosophers and observers have discovered in every portion of our own system, every new step of our knowledge showing us something new in this respect; and if we combine this consideration with the thought how small a portion of the universe our knowledge includes, we shall, without being able at all to discern the extent of the skill and wisdom thus displayed, see something of the character of the design, and of the copiousness and ampleness of the means which the scheme of the world exhibits. And when we see that the tendency of all the arrangements which we can comprehend is to support the existence, to de- velope the faculties, to promote the well-being of these countless species of creatures ; we shall have some impression of the beneficence and love of the Creator, as manifested in the physical government of his creation.

2. It is extremely difficult to devise any means of bringing before a common apprehension the scale on which the universe is constructed, the enormous proportion which the larger dimensions bear to the smaller, and the amazing number of steps from large to smaller, or from small to larger, which the consideration of it offers. The following comparative representations may serve to give the reader to whom the subject is new some idea of these steps.

If we suppose the earth to be represented by a globe a foot in diameter, the distance of the sun from the earth will be about two miles ; the diameter of the sun, on the same supposition, will be something above one hundred feet, and consequently his bulk such as might be made up of two hemispheres, each about the size of the dome of St. Paul's. The moon will be thirty feet from us, and her diameter three inches, about that of a cricket ball. Thus the sun

would much more than occupy all the space
within the moon's orbit. On the same scale,
Jupiter would be above ten miles from the sun,
and Uranus forty. We see then how thinly
scattered through space are the heavenly bodies.
The fixed stars would be at an unknown distance, but, probably, if all distances were thus
diminished, no star would be nearer to such a
one-foot earth, than the moon now is to us.

On such a terrestrial globe the highest mountains would be about 1-80th of an inch high, and
consequently only just distinguishable. We may
imagine therefore how imperceptible would be the
largest animals. The whole organized covering
of such a globe would be quite undiscoverable by
the eye, except perhaps by colour, like the bloom
on a plum.

In order to restore this earth and its inhabitants to their true dimensions, we must
magnify them forty millions of times ; and to preserve the proportions, we must increase equally
the distances of the sun and of the stars from us.
They seem thus to pass off into infinity ; yet
each of them thus removed, has its system of
mechanical and perhaps of organic processes
going on upon its surface.

But the arrangements of organic life which we
can see with the naked eye are few, compared
with those which the microscope detects. We
know that we may magnify objects thousands of

times, and still discover fresh complexities of structure ; if we suppose, therefore, that we increase every particle of matter in our universe in such a proportion, in length, breadth, and thickness, we may conceive that we tend thus to bring before our apprehension a true estimate of the quantity of organized adaptations which are ready to testify the extent of the Creator's power.

3. The other numerical quantities which we have to consider in the phenomena of the universe are on as gigantic a scale as the distances and sizes. By the rotation of the earth on its axis, the parts of the equator move at the rate of a thousand miles an hour, and the portions of the earth's surface which are in our latitude, at about six hundred.   The former velocity is nearly that with which a cannon ball is discharged from the mouth of a gun ; but, large as it is, it is inconsiderable compared with the velocity of the earth in its orbit about the sun.   This latter velocity is sixty-five times the former.   By the rotatory motion of the earth, a point of its surface is carried sometimes forwards and sometimes backwards with regard to the annual progression ; but in consequence of the great predominance of the latter velocity in amount, the former scarcely affects it either way.   And even the latter velocity is inconsiderable compared with that of light ; which comparison, however, we shall not make ; since, according to the theory we have

considered as most probable, the motion of light is not a transfer of matter but of motion from one part of space to another.

The extent of the scale of density of different substances has already been mentioned ; gold is twenty times as heavy as water ; air is eight hundred and thirty times lighter, steam 1-8000 times lighter than water ; the luminiferous ether is incomparably rarer than steam : and this is true of the matter of light, whether we adopt the undulatory theory or any other.

4. The above statements are vast in amount, and almost oppressive to our faculties. They belong to the measurement of the powers which are exerted in the universe, and of the spaces through which their efficacy reaches (for the most distant bodies are probably connected both by gravity and light). But these estimates cannot be said so much to give us any notion of the powers of the Deity, as to correct the errors we should fall into by supposing his powers at all to resemble ours :—by supposing that numbers, and spaces, and forces, and combinations, which would overwhelm us, are any obstacle to the arrangements which his plan requires. We can easily understand that to an intelligence surpassing ours in degree only, that may be easy which is impossible to us. The child who cannot count beyond four, the savage who has no name for any number above five, cannot comprehend the possibility of

dealing with thousands and millions : yet a little additional developement of the intellect makes such numbers manageable and conceivable. The difficulty which appears to reside in numbers and magnitudes and stages of subordination, is one produced by judging from ourselves—by measuring with our own sounding line ; when that reaches no bottom, the ocean appears unfathomable. Yet in fact, how is a hundred millions of miles a *great* distance? how is a hundred millions of times a *great* ratio ? Not in itself: this *greatness* is no quality of the numbers which can be proved like their mathematical properties ; on the contrary, all that absolutely belongs to number, space, and ratio, must, we know demonstrably, be equally true of the largest and the smallest. It is clear that the *greatness* of these expressions of measure has reference to *our* faculties only. Our astonishment and embarrassment take for granted the limits of our own nature. We have a tendency to treat a difference of degree and of addition, as if it were a difference of kind and of transformation. The existence of the attributes, design, power, goodness, is a matter depending on obvious grounds : about these qualities there can be no mistake : if we can know anything, we can know these attributes when we see them. But the extent, the limits of such attributes must be determined by their effects ; our knowledge of their limits by what we see of the effects. Nor is any extent, any amount of power and goodness improbable before hand: we know that these must be great, we cannot tell how great. We should not expect beforehand to find them bounded; and therefore when the boundless prospect opens before us, we may be bewildered, but we have no reason to be shaken in our conviction of the reality of the cause from which their effects proceed: we may feel ourselves incapable of following the train of thought, and may stop, but we have no rational motive for quitting the point which we have thus attained in tracing the Divine Perfections.

On the contrary, those magnitudes and proportions which leave our powers of conception far behind;—that ever-expanding view which is brought before us, of the scale and mechanism, the riches and magnificence, the population and activity of the universe;—may reasonably serve, not to disturb, but to enlarge and elevate our conceptions of the Maker and Master of all; to feed an ever-growing admiration of His wonderful nature; and to excite a desire to be able to contemplate more steadily and conceive less inadequately the scheme of his government and the operation of his power.

Chapter III.

On Man's Place in the Universe.

THE mere aspect of the starry heavens, without taking into account the view of them to which science introduces us, tends strongly to force upon man the impression of his own insignificance. The vault of the sky arched at a vast and unknown distance over our heads ; the stars, apparently infinite in number, each keeping its appointed place and course, and seeming to belong to a wide system of things which has no relation to the earth ; while man is but one among many millions of the earth's inhabitants ; — all this makes the contemplative spectator feel how exceedingly small a portion of the universe he is ; how little he must be, in the eyes of an intelligence which can embrace the whole. Every person, in every age and country, will recognize as irresistibly natural the train of thought expressed by the Hebrew psalmist : " when I consider the heavens the work of thy hands—the moon and the stars which thou hast ordained— Lord what is man that thou art mindful of him, or the son of man that thou regardest him ?"

If this be the feeling of the untaught person,

when he contemplates the aspect of the skies, such as they offer themselves to a casual and unassisted glance, the impression must needs be incalculably augmented, when we look at the universe with the aid of astronomical discovery and theory. We then find, that a few of the shining points which we see scattered on the face of the sky in such profusion, appear to be of the same nature as the earth, and may perhaps, as analogy would suggest, be like the earth, the habitations of organized beings ;—that the rest of " the host of heaven" may, by a like analogy, be conjectured to be the centres of similar systems of revolving worlds ;—that the vision of man has gone travelling onwards, to an extent never anticipated, through this multitude of systems, and that while myriads of new centres start up at every advance, he appears as yet not to have received any intimation of a limit. Every person probably feels, at first, lost, confounded, over-whelmed, with the vastness of this spectacle ; and seems to himself, as it were, annihilated by the magnitude and multitude of the objects which thus compose the universe. The distance between him and the Creator of the world appears to be increased beyond measure by this disclosure. It seems as if a single individual could have no chance and no claim for the regard of the Ruler of the whole.

The mode in which the belief of God s govern-

ment of the physical world is important and interesting to man, is, as has already been said,
through the connexion which this belief has
with the conviction of God's government of the
moral world; this latter government being, from
its nature, one which has a personal relation to
each individual, his actions and thoughts. It
will, therefore, illustrate our subject to show that
this impression of the difficulty of a personal
superintendence and government, exercised by
the Maker of the world over each of his rational
and free creatures, is founded upon illusory
views; and that on an attentive and philosophical examination of the subject, such a government is in accordance with all that we can discover of the scheme and the scale of the universe.

1. We may, in the first place, repeat the
observation made in the last chapter, on the confusion which sometimes arises in our minds, and
makes us consider the number of the objects of
the Divine care as a difficulty in the way of its
exercise. If we can conceive this care employed
on a million persons, on the population of a kingdom, of a city, of a street, there is no real difficulty in supposing it extended to every planet in
the solar system, admitting each to be peopled
as ours is; nor to every part of the universe,
supposing each star the centre of such a system.
Numbers are nothing in themselves; and when

we reject the known, but unessential limits of our own faculties, it is quite as allowable to suppose a million millions of earths, as one, to be under the moral government of God.

2. In the next place we may remark, not only that no reason can be assigned why the Divine care should not extend to a much greater number of individuals than we at first imagine, but that in fact we know that it *does* so extend. It has been well observed, that about the same time when the invention of the telescope showed us that there might be myriads of other worlds claiming the Creator's care; the invention of the microscope proved to us that there were in our own world myriads of creatures, before unknown, which this care was preserving. While one discovery seemed to remove the Divine Providence further from us, the other gave us most striking examples that it was far more active in our neighbourhood than we had supposed: while the first extended the boundaries of God's known kingdom, the second made its known administration more minute and careful. It appeared that in the leaf and in the bud, in solids and in fluids, animals existed hitherto unsuspected; the apparently dead masses and blank spaces of the world were found to swarm with life. And yet, of the animals thus revealed, all, though unknown to us before, had never been forgotten by Providence. Their structure, their vessels and limbs,

their adaptation to their situation, their food and habitations, were regulated in as beautiful and complete a manner as those of the largest and apparently most favoured animals. The smallest insects are as exactly finished, often as gaily ornamented, as the most graceful beasts or the birds of brightest plumage. And when we seem to go out of the domain of the complex animal structure with which we are familiar, and come to animals of apparently more scanty faculties, and less developed powers of enjoyment and action, we still find that their faculties and their senses are in exact harmony with their situation and circumstances; that the wants which they have are provided for, and the powers which they possess called into activity. So that Müller, the patient and accurate observer of the smallest and most obscure microscopical animalcula, declares that all classes alike, those which have manifest organs, and those which have not, offer a vast quantity of new and striking views of the animal economy; every step of our discoveries leading us to admire the design and care of the Creator.* We find, therefore, that the Divine Providence is, in fact, capable of extending itself adequately to an immense succession of tribes of beings, surpassing what we can image or could previously have anticipated; and thus we may

* Müller, Infusoria, Preface.

feel secure, so far as analogy can secure us, that the mere multitude of created objects cannot remove us from the government and superintendence of the Creator.

3. We may observe further, that, vast as are the parts and proportions of the universe, we still appear to be able to perceive that it is *finite;* the subordination of magnitudes and numbers and classes appears to have its limits. Thus, for anything which we can discover, the sun is the largest body in the universe; and at any rate, bodies of the order of the sun are the largest of which we have any evidence: we know of no substance denser than gold, and it is improbable that one denser, or at least much denser, should ever be detected : the largest animals which exist in the sea and on the earth are almost certainly known to us. We may venture also to say, that the smallest animals which possess in their structure a clear analogy with larger ones, have been already seen. Many of the animals which the microscope detects, are as complete and complex in their organization as those of larger size : but beyond a certain point, they appear, as they become more minute, to be reduced to a homogeneity and simplicity of composition which almost excludes them from the domain of animal life. The smallest microscopical objects which can be supposed to be organic, are points,* or

* *Monas.*   Müller.   Cuvier.

gelatinous globules,* or threads,† in which no distinct organs, interior or exterior, can be discovered. These, it is clear, cannot be considered as indicating an indefinite progression of animal life in a descending scale of minuteness. We can, mathematically speaking, conceive one of these animals as perfect and complicated in its structure as an elephant or an eagle, but we do not find it so in nature. It appears, on the contrary, in these objects, as if we were, at a certain point of magnitude, reaching the boundaries of the animal world. We need not here consider the hypotheses and opinions to which these ambiguous objects have given rise ; but, without any theory, they tend to show that the subordination of organic life is finite on the side of the little as well as of the great.

Some persons might, perhaps, imagine that a ground for believing the smallness of organized beings to be limited, might be found in what we know of the constitution of matter. If solids and fluids consist of particles of a definite, though exceeding smallness, which cannot further be divided or diminished, it is manifest that we have, in the smallness of these particles, a limit to the possible size of the vessels and organs of animals. The fluids which are secreted, and which circulate in the body of a mite, must

* *Volvox.*   † *Vibrio.*   Müller. Cuvier.

needs consist of a vast number of particles, or
they would not be fluids : and an animal might
be so much smaller than a mite, that its tubes
could not contain a sufficient collection of the
atoms of matter, to carry on its functions. We
should, therefore, of necessity reach a limit of
minuteness in organic life, if we could demonstrate that matter is composed of such indivisible
atoms. We shall not, however, build anything
on this argument ; because, though the *atomic
theory* is sometimes said to be proved, what is
proved is, that chemical and other effects take
place as if they were the aggregate of the effects
of certain particles of elements, the *proportions*
of which particles are fixed and definite ; but that
any limit can be assigned to the smallness of
these particles, has never yet been made out.
We prefer, therefore, to rest the proof of the
finite extent of animal life, as to size, on the
microscopical observations previously referred to.

Probably we cannot yet be said to have
reached the limit of the universe with the power
of our telescopes ; that is, it does not appear that
telescopes have yet been used, so powerful in
exhibiting small stars, that we can assume that
more powerful instruments would not discover
new stars. Whether or no, however, this degree
of perfection has been reached, we have no proof
that it does not exist ; if it were once obtained
we should have, with some approximation, the

limit of the universe as to the number of worlds, as we have already endeavoured to show we have obtained the limits with regard to the largeness and smallness of the inhabitants of our own world.

In like manner, although the discovery of new species in some of the kingdoms of nature has gone on recently with enormous rapidity, and to an immense extent ;—for instance in botany, where the species known in the time of Linnæus were about 10,000, and are now probably 50,000 ; —there can be no doubt that the number of species and genera is really limited ; and though a great extension of our knowledge is required to reach these limits, it is our ignorance merely, and not their non-existence, which removes them from us.

In the same way it would appear that the universe, so far as it is an object of our knowledge, is finite in other respects also. Now when we have once attained this conviction, all the oppressive apprehension of being overlooked in the government of the universe has no longer any rational source. For in the superintendence of a finite system of things, what is there which can appear difficult or overwhelming to a Being such as we must, from what we know, conceive the Creator to be ? Difficulties arising from space, number, gradation, are such as we can conceive *ourselves* capable of overcoming, merely by an extension of our present faculties. Is it not then easy to imagine that such difficulties must vanish before Him who made us and our faculties? Let it be considered how enormous a proportion the largest work of man bears to the smallest;—the great pyramid to the point of a needle. This comparison does not overwhelm us, because we know that man has made both. Yet the difference between this proportion and that of the sun to the claw of a mite, does not at all correspond to the difference which we must suppose to obtain between the Creator and the creature. It appears then that, if the first flash of that view of the universe which science reveals to us, does sometimes dazzle and bewilder men, a more attentive examination of the prospect, by the light we thus obtain, shows us how unfounded is the despair of our being the objects of Divine Providence, how absurd the persuasion that we have discovered the universe to be too large for its ruler.

4. Another ground of satisfactory reflexion, having the same tendency, is to be found in the admirable order and consistency, the subordination and proportion of parts, which we find to prevail in the universe, as far as our discoveries reach. We have, it may be, a multitude almost innumerable of worlds, but no symptom of crowding, of confusion, of interference. All such defects are avoided by the manner in which these worlds are distributed into systems;—these systems, each

occupying a vast space, but yet disposed at distances before which their own dimensions shrink
into insignificance ;—all governed by one law,
yet this law so concentrating its operation on
each system, that each proceeds as if there were
no other, and so regulating its own effects that
perpetual change produces permanent uniformity.
This is the kind of harmonious relation which we
perceive in that part of the universe, the mechanical part namely, the laws of which are best
known to us.   In other provinces, where our
knowledge is more imperfect, we can see glimpses
of a similar vastness of combination, producing,
by its very nature, completeness of detail.   Any
analogy by which we can extend such views to
the moral world, must be of a very wide and
indefinite kind ; yet the contemplation of this
admirable relation of the arrangements of the
physical creation, and the perfect working of
their laws, is well calculated to give us confidence
in a similar beauty and perfection in the arrangements by which our moral relations are directed,
our higher powers and hopes unfolded. We may
readily believe that there is, in this part of the
creation also, an order, a subordination of some
relations to others, which may remove all difficulty
arising from the vast multitude of moral agents
and actions, and make it possible that the superintendence of the moral world shall be directed
with as exact a tendency to moral good, as that

by which the government of the physical world
is directed to physical good.

We may perhaps see glimpses of such an
order, in the arrangements by which our highest
and most important duties depend upon our
relation to a small circle of persons immediately
around us : and again, in the manner in which
our acting well or ill results from the operation of
a few principles within us ; as our conscience,
our desire of moral excellence, and of the favour
of God. We can hardly consider such principles
otherwise than as intended to occupy their proper
place in the system by which man's destination
is to be determined ; and thus, as among the
means of the government and superintendence of
God in the moral world.

That there must be an order and a system to
which such regulative principles belong, the
whole analogy of creation compels us to believe.
It would be strange indeed, 'if, while the mechanical world, the system of inert matter, is so
arranged that we cannot contemplate its order
without an elevated intellectual pleasure ;—while
organized life has no faculties without their proper
scope, no tendencies without their appointed object ;—the rational faculties and moral tendencies
of man should belong to no systematic order,
should operate with no corresponding purpose :
that, while the perception of sweet and bitter has
its acknowledged and unmistakeable uses, the

universal perception of right and wrong, the unconquerable belief of the merit of certain feelings and actions, the craving alike after moral advancement and after the means of attaining it, should exist only to delude, perplex, and disappoint man. No one, with his contemplations calmed and filled and harmonized by the view of the known constitution of the universe, its machinery " wheeling unshaken" in the farthest skies and in the darkest cavern, its vital spirit breathing alike effectively in the veins of the philosopher and the worm ;—no one, under the influence of such a train of contemplations, can possibly admit into his mind a persuasion which makes the moral part of our nature a collection of inconsistent and futile impressions, of idle dreams and warring opinions, each having the same claims to our acceptance. Wide as is the distance between the material and the moral world ; shadowy as all reasonings necessarily are which attempt to carry the inferences of one into the other ; elevated above the region of matter as all the principles and grounds of truth must be, which belong to our responsibilities and hopes ; still the astronomical and natural philosopher can hardly fail to draw from their studies an imperturbable conviction that our moral nature cannot correspond to those representations according to which it has no law, coherency, or object. The mere natural reasoner may, or

must, stop far short of all that it is his highest
interest to know, his first duty to pursue; but
even he, if he take any elevated and comprehensive views of his own subject, must escape from
the opinions, as unphilosophical as they are
comfortless, which would expel from our view
of the world all reference to duty and moral
good, all reliance on the most universal grounds
of trust and hope.

Men's belief of their duty, and of the reasons
for practising it, connected as it is with the conviction of a personal relation to their Maker, and
of His power of superintendence and reward, is
as manifest a fact in the moral, as any that can
be pointed out is in the natural world. By the
mere analogy which has been intimated, therefore, we cannot but conceive that this fact belongs in some manner or other to the order of the
moral world, and of its government.

When any one acknowledges a moral governor
of the world; perceives that domestic and social
relations are perpetually operating and seem intended to operate, to retain and direct men in the
path of duty; and feels that the voice of conscience, the peace of heart which results from a
course of virtue, and the consolations of devotion,
are ever ready to assume their office as our guides
and aids in the conduct of all our actions;—he
will probably be willing to acknowledge also that
the means of moral government are not wanting,

and will no longer be oppressed or disturbed by
the apprehension that the superintendence of the
world may be too difficult for its Ruler, and that
any of His subjects and servants may be overlooked. He will no more fear that the moral than
that the physical laws of God's creation should be
forgotten in any particular case : and as he knows
that every sparrow which falls to the ground
contains in its structure innumerable marks of
the Divine care and kindness, he will be persuaded that every individual, however apparently
humble and insignificant, will have his moral
being dealt with according to the laws of God's
wisdom and love ; will be enlightened, supported,
and raised, if he use the appointed means which
God's administration of the world of moral light
and good offers to his use.

CHAPTER IV.

*On the Impression produced by the contemplation
of Laws of Nature ; or, on the Conviction that
Law implies Mind.*

THE various trains of thought and reasoning
which lead men from a consideration of the
natural world to the conviction of the existence,
the power, the providence of God, do not require,
for the most part, any long or laboured deduction,

to give them their effect on the mind. On the contrary, they have, in every age and country, produced their impression on multitudes who have not instituted any formal reasonings upon the subject, and probably upon many who have not put their conclusions in the shape of any express propositions. The persuasion of a superior intelligence and will, which manifests itself in every part of the material world, is, as is well known, so widely diffused and deeply infixed, as to have made it a question among speculative men whether the notion of such a power is not universal and innate. It is our business to show only how plainly and how universally such a belief results from the study of the appearances about us. That in many nations, in many periods, this persuasion has been mixed up with much that was erroneous and perverse in the opinions of the intellect or the fictions of the fancy, does not weaken the force of such consent. The belief of a supernatural and presiding power runs through all these errors : and while the perversions are manifestly the work of caprice and illusion, and vanish at the first ray of sober enquiry, the belief itself is substantial and consistent, and grows in strength upon every new examination. It was the firmness and solidity of the conviction of *something* Divine which gave a hold and permanence to the figments of so many false divinities. And those who have traced the progress of human thought on other subjects, will not think it strange, that while the fundamental persuasion of a Deity was thus irremoveably seated in the human mind, the developement of this conception into a consistent, pure, and steadfast belief in one Almighty and Holy Father and God, should be long missed, or never attained, by the struggle of the human faculties; should require long reflexion to mature it, and the aid of revelation to establish it in the world.

The view of the universe which we have principally had occasion to present to the reader, is that in which we consider its appearances as reducible to certain fixed and general laws. Availing ourselves of some of the lights which modern science supplies, we have endeavoured to show that the adaptation of such laws to each other, and their fitness to promote the harmonious and beneficial course of the world, may be traced, wherever we can discover the laws themselves; and that the conceptions of the Divine Power, Goodness and Superintendence which we thus form, agree in a remarkable manner with the views of the Supreme Being, to which reason, enlightened by the divine revelation, has led.

But we conceive that the general impressions of mankind would go further than a mere assent to the argument as we have thus stated it. To most persons it appears that the mere existence of a

law connecting and governing any class of phenomena, implies a presiding intelligence which
has preconceived and established the law. When
events are regulated by precise rules of time and
space, of number and measure, men conceive
these rules to be the evidence of thought and
mind, even without discovering in the rules any
peculiar adaptations, or without supposing their
purpose to be known.

The origin and the validity of such an impression on the human mind may appear to some
matters of abstruse and doubtful speculation : yet
the tendency to such a belief prevails strongly and
widely, both among the common class of minds
whose thoughts are casually and unsystematically
turned to such subjects, and among philosophers
to whom laws of nature are habitual subjects of
contemplation. We conceive therefore that such
a tendency may deserve to be briefly illustrated ;
and we trust also that some attention to this point
may be of service in throwing light upon the true
relation of the study of nature to the belief in
God.

1. A very slight attention shows us how readily
order and regularity suggest to a common apprehension the operation of a calm and untroubled
intelligence presiding over the course of events.
Thus the materialist poet, in accounting for the
belief in the Gods, though he does not share it,

cannot deny the habitual effect of this manifestation.

> Præterea cœli rationes *ordine certo*
> Et *varia* annorum cernebant *tempora* vorti ;
> Nec poterant quibus id fieret cognoscere caussis.
> Ergo perfugium sibi habebant omnia Divis
> Tradere et illorum nutu facere omnia flecti.
>
> LUCRET. v. 1182.

> They saw the skies in constant order run,
> The varied seasons and the circling sun,
> Apparent rule, with unapparent cause,
> And thus they sought in Gods the source of laws.

The same feeling may be traced in the early mythology of a large portion of the globe. We might easily, taking advantage of the labours of learned men, exemplify this in the case of the oriental nations, of Greece, and of many other countries. Nor does there appear much difficulty in pointing out the error of those who have maintained that all religion had its *origin* in the worship of the stars and the elements ; and who have insinuated that all such impressions are unfounded, inasmuch as these are certainly not right objects of human worship. The religious feeling, the conviction of a supernatural power, of an intelligence connecting and directing the phenomena of the world, had not its *origin* in the worship of sun, or stars, or elements ; but was itself the necessary though unexpressed foundation of all worship, and all forms of false,

as well as true, religion. The contemplation of
the earth and heavens called into action this
religious tendency in man; and to say that the
worship of the material world formed or suggested
this religious feeling, is to invert the order of
possible things in the most unphilosophical manner. Idolatry is not the source of the belief in
God, but is a compound of the persuasion of a
supernatural government, with certain extravagant and baseless conceptions as to the manner
in which this government is exercised.

We will quote a passage from an author who
has illustrated at considerable length the hypothesis that all religious belief is derived from the
worship of the elements.

" Light, and darkness its perpetual contrast;
the succession of days and nights, the periodical
order of the seasons; the career of the brilliant
luminary which regulates their course; that of
the moon his sister and rival; night, and the
innumerable fires which she lights in the blue
vault of heaven; the revolutions of the stars,
which exhibit them for a longer or a shorter
period above our horizon; the constancy of this
period in the fixed stars, its variety in the wandering stars, the planets; their direct and retro-
gade course, their momentary rest; the phases
of the moon waxing, full, waning, divested of all
light; the progressive motion of the sun upwards,
downwards : the successive order of the rising

and setting of the fixed stars which mark the different points of the course of the sun, while the various aspects which the earth itself assumes mark, here below also, the same periods of the sun's annual motion ; . . . all these different pictures, displayed before the eyes of man, formed the great and magnificent spectacle by which I suppose him surrounded at the moment when *he is about to create his gods.*"[*]

What is this (divested of its wanton levity of expression) but to say, that when man has so far traced the course of nature as to be irresistibly impressed with the existence of order, law, variety in constancy, and fixity in change ; of relations of form and space, duration and succession, cause and consequence, among the objects which surround him ; there springs up in his breast, unbidden and irresistibly, the thought of superintending intelligence, of a mind which comprehended from the first and completely that which he late and partially comes to know ? The worship of earth and sky, of the host of heaven and the influences of nature, is not the ultimate and fundamental fact in the early history of the religious impressions of mankind. These are but derivative streams, impure and scanty, from the fountain of religious feeling which appears to be disclosed to us by the contemplation of the uni-

* Dupuis. Origine des Cultes.

verse, as the seat of law and the manifestation of intellect. Time suggests to man the thought of eternity ; space of infinity ; law of intelligence ; order of purpose ; and however difficult and long a task it may be to develope these suggestions into clear convictions, these thoughts are the real parents of our natural religious belief. The only relation between true religion and the worship of the elemental world is, that the latter is the partial and gross perversion, the former the consistent and pure developement of the same original idea.

2. The connexion of the laws of the material world with an intelligence which preconceived and instituted the law, which is thus, as we perceive, so generally impressed on the common apprehension of mankind, has also struck no less those who have studied nature with a more systematic attention, and with the peculiar views which belong to science. The laws which such persons learn and study, seem, indeed, most naturally to lead to the conviction of an intelligence which originally gave to the law its form.

What we call a general law is, in truth, a form of expression including a number of facts of like kind. The facts are separate ; the unity of view by which we associate them, the character of generality and of law, resides in those relations which are the object of the intellect. The law once apprehended by us, takes in our minds the place of the facts themselves, and is said to *govern* or determine them, because it determines our anticipations of what they will be. But we cannot, it would seem, conceive a law, founded on such intelligible relations, to govern and determine the facts themselves, any otherwise than by supposing also an intelligence by which these relations are contemplated, and these consequences realised. We cannot then represent to ourselves the universe governed by general laws otherwise than by conceiving an intelligent and conscious Deity, by whom these laws were originally contemplated, established, and applied.

This perhaps will appear more clear, when it is considered that the laws of which we speak are often of an abstruse and complex kind, depending upon relations of space, time, number, and other properties, which we perceive by great attention and thought. These relations are often combined so variously and curiously, that the most subtle reasonings and calculations which we can form are requisite in order to trace their results. Can such laws be conceived to be instituted without any exercise of knowledge and intelligence? can material objects apply geometry and calculation to themselves? can the lenses of the eye, for instance, be formed and adjusted with an exact suitableness to their refractive powers, while there is in the agency which has framed them, no consciousness of the laws of light, of the course of rays, of the visible properties of things? This appears to be altogether inconceivable.

Every particle of matter possesses an almost endless train of properties, each acting according to its peculiar and fixed laws. For every atom of the same kind of matter these laws are invariably and perpetually the same, while for different kinds of matter the difference of these properties is equally constant. This constant and precise resemblance, this variation equally constant and equally regular, suggest irresistibly the conception of some cause, independent of the atoms themselves, by which their similarity and dissimilarity, the agreement and difference of their deportment under the same circumstances, have been determined. Such a view of the constitution of matter, as is observed by an eminent writer of our own time, effectually destroys the idea of its eternal and self-existent nature, " by giving to each of its atoms the essential characters, at once, of a *manufactured article* and a *subordinate agent*."*

That such an impression, and the consequent belief in a divine Author of the universe, by whom its laws were ordained and established, does result from the philosophical contemplation of nature, will, we trust, become still more evident by tracing the effect produced upon men's minds

* Herschel on the Study of Nat. Phil. Art. 28

by the discovery of such laws and properties as
those of which we have been speaking ; and
we shall therefore make a few observations on
this subject.

Chapter V.

*On Inductive Habits ; or, on the Impression pro-
duced on Men's Minds by discovering Laws of
Nature.*

THE object of physical science is to discover such
laws and properties as those of which we have
spoken in the last chapter. In this task, undoubtedly a progress has been made on which
we may well look with pleasure and admiration ;
yet we cannot hesitate to confess that the extent
of our knowledge on such subjects bears no proportion to that of our ignorance. Of the great
and comprehensive laws which rule over the
widest provinces of natural phenomena, few have
yet been disclosed to us. And the names of the
philosophers, whose high office it has been to
detect such laws, are even yet far from numerous.
In looking back at the path by which science
has advanced to its present position, we see the
names of the great discoverers shine out like
luminaries, few and scattered along the line : by

far the largest portion of the space is occupied
by those whose comparatively humble office it
was to verify, to develope, to apply the general
truths which the discoverers brought to light.

It will readily be conceived that it is no easy
matter, if it be possible, to analyse the process of
thought by which laws of nature have thus been
discovered ; a process which, as we have said,
has been in so few instances successfully performed. We shall not here make any attempt
at such an analysis. But without this, we conceive it may be shown that the constitution and
employment of the mind on which such discoveries depend, are friendly to that belief in a
wise and good Creator and Governor of the
world, which it has been our object to illustrate
and confirm. And if it should appear that those
who see further than their fellows into the bearings and dependencies of the material things and
elements by which they are surrounded, have
also been, in almost every case, earnest and forward in acknowledging the relation of all things
to a supreme intelligence and will ; we shall be
fortified in our persuasion that the true scientific
perception of the general constitution of the
universe, and of the mode in which events are
produced and connected, is fitted to lead us to
the conception and belief of God.

Let us consider for a moment what takes place
in the mind of a student of nature when he

attains to the perception of a law previously unknown, connecting the appearances which he
has studied. A mass of facts which before
seemed incoherent and unmeaning, assume, on a
sudden, the aspect of connexion and intelligible
order. Thus, when Kepler discovered the law
which connects the periodic times with the
diameters of the planetary orbits ; or, when
Newton showed how this and all other known
mathematical properties of the solar system were
included in the law of universal gravitation according to the inverse square of the distance ;
particular circumstances which, before, were
merely matter of independent record, became,
from that time, indissolubly conjoined by the laws
so discovered. The separate occurrences and
facts, which might hitherto have seemed casual
and without reason, were now seen to be all exemplifications of the same truth. The change is
like that which takes place when we attempt to
read a sentence written in difficult or imperfect
characters. For a time the separate parts appear to be disjointed and arbitrary marks ; the
suggestions of possible meanings, which succeed
each other in the mind, fail, as fast as they are
tried, in combining or accounting for these sym-
bols : but at last the true supposition occurs ;
some words are found to coincide with the meaning thus assumed ; the whole line of letters appear to take definite shapes and to leap into

their proper places ; and the truth of the happy conjecture seems to flash upon us from every part of the inscription.

The discovery of laws of nature, truly and satisfactorily connecting and explaining phenomena, of which, before, the connexion and causes had been unknown, displays much of a similar process, of obscurity succeeded by evidence, of effort and perplexity followed by conviction and repose. The innumerable conjectures and failures, the glimpses of light perpetually opening and as often clouded over, the unwearied perseverance and inexhaustible ingenuity exercised by Kepler in seeking for the laws which he finally discovered, are, thanks to his communicative disposition, curiously exhibited in his works, and have been narrated by his biographers ; and such efforts and alternations, modified by character and circumstances, must generally precede the detection of any of the wider laws and dependencies by which the events of the universe are regulated. We may readily conceive the satisfaction and delight with which, after this perplexity and struggle, the discoverer finds himself in light and tranquillity ; able to look at the province of nature which has been the subject of his study, and to read there an intelligible connexion, a sufficing reason, which no one before him had understood or apprehended.

This step so much resembles the mode in which one intelligent being understands and apprehends the conceptions of another, that we cannot be surprised if those persons in whose minds such a process has taken place, have been most ready to acknowledge the existence and operation of a superintending intelligence, whose ordinances it was their employment to study. When they had just read a sentence of the table of the laws of the universe, they could not doubt whether it had had a legislator. When they had decyphered there a comprehensive and substantial truth, they could not believe that the letters had been thrown together by chance. They could not but readily acknowledge that what their faculties had enabled them to read, must have been written by some higher and profounder mind. And accordingly, we conceive it will be found, on examining the works of those to whom we owe our knowledge of the laws of nature, and especially of the wider and more comprehensive laws, that such persons have been strongly and habitually impressed with the persuasion of a Divine Purpose and Power which had regulated the events which they had attended to, and ordained the laws which they had detected.

To those who have pursued science without reaching the rank of discoverers;—who have possessed a derivative knowledge of the laws of nature which others had disclosed, and have em-

ployed themselves in tracing the consequences of
such laws, and systematising the body of truth
thus produced, the above description does not
apply ; and we have not therefore in these cases
the same ground for anticipating the same frame
of mind. If among men of science of this class, the
persuasion of a supreme Intelligence has at some
periods been less vivid and less universal, than in
that higher class of which we have before spoken,
the fact, so far as it has existed, may perhaps be
in some degree accounted for.   But whether the
view which we have to give of the mental peculiarities of men whose science is of this derivative
kind be well founded, and whether the account
we have above offered of that which takes place
in the minds of original discoverers of laws in
scientific researches be true, or not, it will probably be considered a matter of some interest to
point out historically that in fact, such discoverers
have been peculiarly in the habit of considering
the world as the work of God.   This we shall
now proceed to do.

As we have already said, the names of *great*
discoverers are not very numerous.   The sciences
which we may look upon as having reached or at
least approached their complete and finished
form, are Mechanics, Hydrostatics, and Physical
Astronomy.   Galileo is the father of modern
Mechanics ; Copernicus, Kepler, and Newton
are the great names which mark the progress

of Astronomy. Hydrostatics shared in a great measure the fortunes of the related science of Mechanics ; Boyle and Pascal were the persons mainly active in developing its more peculiar principles. The other branches of knowledge which belong to natural philosophy, as Chemistry and Meteorology, are as yet imperfect, and perhaps infant sciences ; and it would be rash to presume to select, in them, names of equal preeminence with those above mentioned : but it may not be difficult to show, with sufficient evidence, that the effect of science upon the authors of science is, in these subjects as in the former ones, far other than to alienate their minds from religious trains of thought, and a habit of considering the world as the work of God.

We shall not dwell much on the first of the above mentioned great names, Galileo ; for his scientific merit consisted rather in adopting the sound philosophy of others, as in the case of the Copernican system, and in combating prevalent errors, as in the case of the Aristotelian doctrines concerning motion, than in any marked and prominent discovery of new principles. Moreover the mechanical laws which he had a share in bringing to light, depending as they did, rather on detached experiments and transient facts, than on observation of the general course of the universe, could not so clearly suggest any reflexion on the government of the world at that period, as they did afterwards when Newton showed their bearing on the cosmical system. Yet Galileo, as a man of philosophical and inventive mind, who produced a great effect on the progress of physical knowledge, is a person whose opinions must naturally interest us, engaged in our present course of reasoning. There is in his writings little which bears upon religious views, as there is in the nature of his works little to lead him to such subjects. Yet strong expressions of piety are not wanting, both in his letters, and in his published treatises. The persecution which he underwent, on account of his writings in favour of the Copernican system, was grounded, not on his opposition to the general truths of natural religion, which is our main concern at present, nor even on any supposed rejection of any articles of Christian faith, but on the alleged discrepancy between his adopted astronomical views and the declarations of scripture. Some of his remarks may interest the reader.

In his third dialogue on the Copernican system he has occasion to speak of the opinion which holds all parts of the world to be framed for man's use alone : and to this he says, " I would that we should not so shorten the arm of God in the government of human affairs ; but that we should rest in this, that we are certain that God and nature are so occupied in the government of human affairs, that they could not more attend to us if they were charged with the care of the human race alone.'' In the same spirit, when some objected to the asserted smallness of the Medicean stars, or satellites of Jupiter, and urged this as a reason why they were unworthy the regard of philosophers, he replied that they are the works of God's power, the objects of His care, and therefore may well be considered as sublime subjects for man's study.

In the Dialogues on Mechanics, there occur those observations concerning the use of the air-bladder in fishes, and concerning the adaptation of the size of animals to the strength of the materials of which they are framed, which have often since been adopted by writers on the wisdom of Providence. The last of the dialogues on the system of the world is closed by a religious reflexion, put in the mouth of the interlocutor who usually expresses Galileo's own opinions. ".While it is permitted us to speculate concerning the constitution of the world, we are also taught (perhaps in order that the activity of the human mind may not pause or languish) that our powers do not enable us to comprehend the works of His hands. May success therefore attend this intellectual exercise, thus permitted and appointed for us ; by which we recognize and admire the greatness of God the more, in proportion as we find ourselves the less able to penetrate the profound abysses of his wisdom.'' And that this train of thought was habitual to the philosopher we have abundant evidence in many other parts of his writings. He had already said in the same dialogue, " Nature (or God, as he elsewhere speaks) employs means in an admirable and inconceivable manner ; admirable, that is, and inconceivable to us, but not to her, who brings about with consummate facility and simplicity things which affect our intellect with infinite astonishment. That which is to us most difficult to understand is to her most easy to execute."

The establishment of the Copernican and Newtonian views of the motions of the solar system and their causes, were probably the occasions on which religious but unphilosophical men entertained the strongest apprehensions that the belief in the government of God may be weakened when we thus " thrust some mechanic cause into his place." It is therefore fortunate that we can show, not only that this ought not to occur, from the reason of the thing, but also that in fact the persons who are the leading characters in the progress of these opinions were men of clear and fervent piety.

In the case of Copernicus himself it does not appear that, originally, any apprehensions were entertained of any dangerous discrepancy between his doctrines and the truths of religion, either natural or revealed. The work which contains these memorable discoveries was addressed to

Pope Paul III., the head, at that time, (1543) of the religious world; and was published, as the author states in the preface, at the urgent entreaty of friends, one of whom was a cardinal, and another a bishop.* "I know," he says, "that the thoughts of a philosopher are far removed from the judgment of the vulgar; since it is his study to search out truth in all things, as far as that is permitted by God to human reason." And though the doctrines are for the most part stated as portions of a mathematical calculation, the explanation of the arrangement by which the sun is placed in the centre of the system is accompanied by a natural reflexion of a religious cast; "Who in this fair temple would place this lamp in any other or better place than there whence it may illuminate the whole? We find then under this ordination an admirable symmetry of the world, and a certain harmonious connexion of the motion and magnitude of the orbs, such as in any other way cannot be found. Thus the progressions and regressions of the planets all arise from the same cause, the motion of the earth. And that no such movements are seen in the fixed stars, argues their immense distance from us,

* Amici me cunctantem atque etiam reluctantem, retraxerunt, inter quos primus fuit Nicolaus Schonbergius, Cardinalis Capuanus, in omni genere literatum celebris; proximus ille vir mei amantissimus Tidemannus Gisius, episcopus Culmensis, sacrarum ut est et omnium bonarum literarum studiosissimus.—*De Revolutionibus. Præf. ad Paulum III.*

which causes the apparent magnitude of the earth's annual course to become evanescent. So great, in short, is this divine fabric of the great and good God ;"* " this best and most regular artificer of the universe," as he elsewhere speaks.

Kepler was the person, who by further studying " the connexion of the motions and magnitude of the orbs," to which Copernicus had thus drawn the attention of astronomers, detected the laws of this connexion, and prepared the way for the discovery, by Newton, of the mechanical laws and causes of such motions. Kepler was a man of strong and lively piety ; and the exhortation which he addresses to his reader before entering on the exposition of some of his discoveries, may be quoted not only for its earnestness but its reasonableness also.—" I beseech my reader, that not unmindful of the divine goodness bestowed on man, he do with me praise and celebrate the wisdom and greatness of the Creator, which I open to him from a more inward explication of the form of the world, from a searching of causes, from a detection of the errors of vision : and that thus, not only in the firmness and stability of the earth he perceive with gratitude the preservation of all living things in nature as the gift of God, but also that in its motion, so recondite, so admirable, he acknowledge the wisdom of the

* Lib. i. cx.

Creator. But him who is too dull to receive this science, or too weak to believe the Copernican system without harm to his piety, him, I say, I advise that, leaving the school of astronomy, and condemning, if he please, any doctrines of the philosophers, he follow his own path, and desist from this wandering through the universe, and lifting up his natural eyes, with which alone he can see, pour himself out from his own heart in praise of God the Creator ; being certain that he gives no less worship to God than the astronomer, to whom God has given to see more clearly with his inward eye, and who, for what he has himself discovered, both can and will glorify God."

The next great step in our knowledge of the universe, the discovery of the mechanical causes by which its motions are produced, and of their laws, has in modern times sometimes been supposed, both by the friends of religion and by others, to be unfavourable to the impression of an intelligent first cause. That such a supposition is founded in error we have offered what appear to us insurmountable reasons for believing. That in the mind of the great discoverer of this mechanical cause, Newton, the impression of a creating and presiding Deity was confirmed, not shaken, by all his discoveries, is so well known that it is almost superfluous to insist upon the fact. His views of the tendency of science invested it with no dangers of this kind. " The business of natural philosophy is," he says, (Optics, Qu. 28.) " to argue from phenomena without eigning hypotheses, and to deduce causes from effects, till we come to the very first cause, which certainly is not mechanical." " Though every true step made in this philosophy brings us not immediately to the knowledge of the first cause, yet it brings us nearer to it and is on that account highly to be valued." The Scholium, or note, which concludes his great work, the Principia, is a well known and most striking evidence on this point, " This beautiful system of sun, planets, and comets, could have its origin in no other way than by the purpose and command of an intelligent and powerful Being. He governs all things, not as the soul of the world, but as the lord of the universe. He is not only God, but Lord or Governor. We know him only by his properties and attributes, by the wise and admirable structure of things around us, and by their final causes ; we admire him on account of his perfections, we venerate and worship him on account of his government."

Without making any further quotations, it must be evident to the reader that the succession of great philosophers through whom mankind have been led to the knowledge of the greatest of scientific truths, the law of universal gravitation, did, for their parts, see the truths which they disclosed to men in such a light that their religious feelings, their reference of the world to an intelligent Creator and Preserver, their admiration of his attributes, were exalted rather than impaired by the insight which they obtained into the structure of the universe.

Having shown this with regard to the most perfect portion of human knowledge, our knowledge of the motions of the solar system, we shall adduce a few other passages, illustrating the prevalence of the same fact in other departments of experimental science ; although, for reasons which have already been intimated, we conceive that sciences of experiment do not conduct so obviously as sciences of observation to the impression of a Divine Legislator of the material world.

The science of Hydrostatics was constructed in a great measure by the founders of the sister science of Mechanics. Of those who were employed in experimentally establishing the principles peculiarly belonging to the doctrine of fluids, Pascal and Boyle are two of the most eminent names. That these two great philosophers were not only religious, but both of them remarkable for their fervent and pervading devotion, is too well known to be dwelt on. With regard to Pascal, however, we ought not perhaps to pass over an opinion of his, that the existence of God cannot be proved from the external world. " I do not undertake to prove this," says he,

" not only because I do not feel myself sufficiently strong to find in nature that which shall convince obstinate atheists, but because such knowledge without Jesus Christ is useless and sterile." It is obvious that such a state of mind would prevent this writer from encouraging or dwelling upon the grounds of natural religion; while yet he himself is an example of that which we wish to illustrate, that those who have obtained the furthest insight into nature, have been in all ages firm believers in God. " Nature," he says in another place, " has perfections in order to show that she is the image of God, and defects in order to show that she is only his image."*

Boyle was not only a most pious man as well as a great philosopher, but he exerted himself very often and earnestly in his writings to show the bearing of his natural philosophy upon his views of the Divine attributes, and of the government of the world. Many of these dissertations convey trains of thought and reasoning which have never been surpast for their combination of judicious sobriety in not pressing his arguments too far, with fervent devotion in his conceptions of the Divine nature. As examples of these merits, we might adduce almost any portion of his tracts on these subjects; for instance, his " Inquiry into the Final Causes of Natural

* Pensées, Art. viii. 1.

Things ;" his " Free Inquiry into the Vulgar
Notion of Nature ;" his " Christian Virtuoso ;"
and his Essay entitled " The High Veneration
Man's Intellect owes to God." It would be
superfluous to quote at any length from these
works. We may observe, however, that he
notices that general fact which we are at present
employed in exemplifying, that " in almost all
ages and countries the generality of philosophers
and contemplative men were persuaded of the
existence of a Deity from the consideration of the
phenomena of the universe ; whose fabric and
conduct they rationally concluded could not justly
be ascribed either to chance or to any other cause
than a Divine Being." And in speaking of the
religious uses of science, he says : " Though I
am willing to grant that some impressions of
God's wisdom are so conspicuous that even a
superficial philosopher may thence infer that the
author of such works must be a wise agent ; yet
how wise an agent he has in these works expressed himself to be, none but an experimental
philosopher can well discern. And 'tis not by a
slight survey, but by a diligent and skilful
scrutiny, of the works of God, that a man must
be, by a rational and affective conviction, engaged
to acknowledge that the author of nature ' is
wonderful in counsel, and excellent in working.' "

After the mechanical properties of fluids, the
laws of the operation of the chemical and physi-

cal properties of the elements about us, offer themselves to our notice. The relations of heat and of moisture in particular, which play so important a part, as we have seen, in the economy of our world, have been the subject of various researches ; and they have led to views of the operation of such agents, some of which we have endeavoured to present to the reader, and to point out the remarkable arrangements by which their beneficial operation is carried on. That the discoverers of the laws by which such operations are regulated, were not insensible to the persuasion of a Divine care and contrivance which those arrangements suggest, is what we should expect, in agreement with what we have already said, and it is what we find. Among the names of the philosophers to whom we owe our knowledge on these subjects, there are none greater than those of Black, the discoverer of the laws of latent heat, and Dalton, who first gave us a true view of the mode in which watery vapour exists and operates in the atmosphere. With regard to the former of these philosophers, we shall quote Dr. Thomson's account of the views which the laws of latent heat suggested to the discoverer.* " Dr. Black quickly perceived the vast importance of this discovery, and took a pleasure in laying before his students a view of

* Thomson's Hist. of Chemistry, vol. i. 321.

the beneficial effects of this habitude of heat in the economy of nature. During the summer season a vast magazine of heat is accumulated in the water, which by gradually emerging during congelation serves to temper the cold of winter. Were it not for this accumulation of heat in water and other bodies, the sun would no sooner go a few degrees to the south of the equator than we should feel all the horrors of winter."

In the same spirit are Mr. Dalton's reflexions, after pointing out the laws which regulate the balance of evaporation and rain,* which he himself first clearly explained. " It is scarcely possible," says he, " to contemplate without admiration the beautiful system of nature by which the surface of the earth is continually supplied with water, and that unceasing circulation of a fluid so essentially necessary to the very being of the animal and vegetable kingdom takes place."

Such impressions appear thus to rise irresistibly in the breasts of men, when they obtain a sight, for the first time, of the varied play and comprehensive connexions of the laws by which the business of the material world is carried on and its occurrences brought to pass. To dwell upon or develope such reflexions is not here our business. Their general prevalence in the minds

* Manch. Mem. vol. v. p. 346

of those to whom these first views of new truths
are granted, has been, we trust, sufficiently
illustrated.   Nor are the names adduced above,
distinguished as they are, brought forwards as
authorities merely.   We do not claim for the
greatest discoverers in the realms of science any
immunity from error.   In their general opinions
they may, as others may, judge or reason ill.
The articles of their religious belief may be as
easily and as widely as of other men's, imperfect,
perverted, unprofitable.   But on this one point,
the tendency of our advances in scientific knowledge of the universe to lead us up to a belief in
a most wise maker and master of the universe,
we conceive that they who make these advances,
and who feel, as an original impression, that
which others feel only by receiving their teach-
ing, must be looked to with a peculiar attention
and respect.   And what their impressions have
commonly been, we have thus endeavoured to
show.

Chapter VI.

On Deductive Habits; or, on the Impression produced on Men's Minds by tracing the consequences of ascertained Laws.

The opinion illustrated in the last chapter, that the advances which men make in science tend to impress upon them the reality of the Divine government of the world, has often been controverted. Complaints have been made, and especially of late years, that the growth of piety has not always been commensurate with the growth of knowledge, in the minds of those who make nature their study. Views of an irreligious character have been entertained, it is sometimes said, by persons eminently well instructed in all the discoveries of modern times, no less than by the superficial and ignorant. Those who have been supposed to deny or to doubt the existence, the providence, the attributes of God, have in many cases been men of considerable eminence and celebrity for their attainments in science. The opinion that this is the case, appears to be extensively diffused, and this persuasion has probably often produced inquietude and grief in the breasts of pious and benevolent men.

This opinion, concerning the want of religious

convictions among those who have made natural philosophy their leading pursuit, has probably gone far beyond the limits of the real fact. But if we allow that there are any strong cases to countenance such an opinion, it may be worth our while to consider how far they admit of any satisfactory explanation. The fact appears at first sight to be at variance with the view we have given of the impression produced by scientific discovery; and it is moreover always a matter of uneasiness and regret, to have men of eminent talents and knowledge opposed to doctrines which we consider as important truths.

We conceive that an explanation of such cases, if they should occur, may be found in a very curious and important circumstance belonging to the process by which our physical sciences are formed. The first discovery of new general truths, and the developement of these truths when once obtained, are two operations extremely different; imply different mental habits, and may easily be associated with different views and convictions on points out of the reach of scientific demonstration. There would therefore be nothing surprising, or inconsistent with what we have maintained above, if it should appear that while original discoverers of laws of nature are peculiarly led, as we have seen, to believe the existence of a supreme intelligence and purpose; the far greater number of cultivators of science, whose employment it is to learn from others these general laws, and to trace, combine, and apply their consequences, should have no clearness of conviction or security from error on this subject, beyond what belongs to persons of any other class.

This will, perhaps, become somewhat more evident by considering a little more closely the distinction of the two operations of discovery and developement, of which we have spoken above, and the tendency which the habitual prosecution of them may be expected to produce in the thoughts and views of the student.

We have already endeavoured in some measure to describe that which takes place when a new law of nature is discovered. A number of facts in which, before, order and connexion did not appear at all, or appeared by partial and contradictory glimpses, are brought into a point of view in which order and connexion become their essential character. It is seen that each fact is but a different manifestation of the same principle; that each particular is that which it is, in virtue of the same general truth. The inscription is decyphered; the enigma is guessed; the principle is understood; the truth is enunciated.

When this step is once made, it becomes possible to deduce from the truth thus established, a train of consequences often in no small degree long and complex. The process of making these inferences may properly be described by the word Deduction, while the very different process by which a new principle is collected from an assemblage of facts, has been termed Induction ; the truths so obtained and their consequences constitute the results of the Inductive Philosophy ; which is frequently and rightly described as a science which ascends from particular facts to general principles, and then descends again from these general principles to particular applications and exemplifications.

While the great and important labours by which science is really advanced consist in the successive steps of the *inductive* ascent, in the discovery of new laws perpetually more and more general ; by far the greater part of our books of physical science unavoidably consists in *deductive* reasoning, exhibiting the consequences and applications of the laws which have been discovered ; and the greater part of writers upon science have their minds employed in this process of deduction and application.

This is true of many of those who are considered, and justly, as distinguished and profound philosophers.   In the mechanical philosophy, that science which applies the properties of matter and the laws of motion to the explanation of the phenomena of the world, this is peculiarly the case.   The laws, when once discovered, occupy little room in their statement, and when no longer contested, are not felt to need a lengthened proof. But their consequences require far more room and far more intellectual labour. If we take, for example, the laws of motion and the law of universal gravitation, we can express in a few lines, that which, when developed, represents and explains an innumerable mass of natural pheno-mena. But here the course of developement is necessarily so long, the reasoning contains so many steps, the considerations on which it rests are so minute and refined, the complication of cases and of consequences is so vast, and even the involution arising from the properties of space and number so serious, that the most consummate subtlety, the most active invention, the most tenacious power of inference, the widest spirit of combination, must be tasked and tasked severely, in order to solve the problems which belong to this portion of science. And the persons who have been employed on these problems, and who have brought to them the high and admirable qualities which such an office requires, have justly excited in a very eminent degree the ad-miration which mankind feel for great intellectual powers. Their names occupy a distinguished place in literary history ; and probably there are no scientific reputations of the last century higher, and none more merited, than those earned by the great mathematicians who have laboured with such wonderful success in unfolding the mechanism of the heavens ; such for instance as D'Alembert, Clairault, Euler, Lagrange, Laplace.

But it is still important to recollect, that the mental employments of men, while they are occupied in this portion of the task of the formation of science, are altogether different from that which takes place in the mind of a discoverer, who, for the first time, seizes the principle which connects phenomena before unexplained, and thus adds another original truth to our knowledge of the universe. In explaining, as the great mathematicians just mentioned have done, the phenomena of the solar system by means of the law of universal gravitation, the conclusions at which they arrived were really included in the truth of the law itself, whatever skill and sagacity it might require to develope and extricate them from the general principle. But when Newton conceived and established the law itself, he added to our knowledge something which was not contained in any truth previously known, nor deducible from it by any course of mere reasoning. And the same distinction, in all other cases, obtains, between these processes which establish the principles, generally few and simple, on which our sciences rest, and those reasonings and calculations, founded on the principles thus obtained, which constitute by far the larger portion of the common treatises on the most complete of the sciences now cultivated.

Since the difference is so great between the process of inductive generalization of physical facts, and that of mathematical deduction of consequences, it is not surprising that the two processes should imply different mental powers and habits. However rare the mathematical talent, in its highest excellence, may be, it is far more common, if we are to judge from the history of science, than the genius which divines the general laws of nature. We have several good mathematicians in every age ; we have few great discoverers in the whole history of our species.

The distinction being thus clearly established between original discovery and derivative speculation, between the ascent to principles and the descent from them, we have further to observe, that the habitual and exclusive prosecution of the latter process may sometimes exercise an unfavourable effect on the mind of the student, and may make him less fitted and ready to apprehend and accept truths different from those with which his reasonings are concerned. We conceive, for example, that a person labours under gross error, who believes the phenomena of the world to be altogether produced by mechanical causes, and who excludes from his view all reference to an intelligent First Cause and Governor. But we conceive that reasons may be shown which make it more probable that error of such a kind should find a place in the mind of a person of deductive, than of inductive habits ;—of a mere mathematician or logician, than of one who studies the facts of the natural world and detects their laws.

The person whose mind is employed in reducing to law and order and intelligible cause the complex facts of the material world, is compelled to look beyond the present state of his knowledge, and to turn his thoughts to the existence of principles higher than those which he yet possesses. He has seen occasions when facts that at first seemed incoherent and anomalous, were reduced to rule and connexion ; and when limited rules were discovered to be included in some rule of superior generality. He knows that all facts and appearances, all partial laws, however confused and casual they at present seem, must still, in reality, have this same kind of bearing and dependence ;—must be bound together by some undiscovered principle of order ; —must proceed from some cause working by most steady rules ;—must be included in some wide and fruitful general truth. He cannot therefore consider any principles which he has already obtained, as the ultimate and sufficient reason of that which he sees. There must be some higher principle, some ulterior reason. The effort and struggle by which he endeavours to extend his view, makes him feel that there is a region of truth not included in his present physical know-

ledge; the very imperfection of the light in
which he works his way, suggests to him that
there must be a source of clearer illumination at
a distance from him.

We must allow that it is scarcely possible to
describe in a manner free from some vagueness
and obscurity, the effect thus produced upon the
mind by the efforts which it makes to reduce
natural phenomena to general laws. But we
trust it will still be allowed that there is no difficulty in seeing clearly that a different influence
may result from this process, and from the process of deductive reasoning which forms the
main employment of the mathematical cultivators
and systematic expositors of physical science in
modern times. Such persons are not led by
their pursuits to anything beyond the general
principles, which form the basis of their explanations and applications. They acquiesce in these;
they make these their ultimate grounds of truth;
and they are entirely employed in unfolding the
particular truths which are involved in the general truth. Their thoughts dwell little upon the
possibility of the laws of nature being other than
we find them to be, or on the reasons why they
are not so; and still less on those facts and
phenomena which philosophers have not yet
reduced to any rule; which are lawless to us,
though we know that, in reality, they are governed
by some principle of order and harmony. On

the contrary, by assuming perpetually the existing laws as the basis of their reasoning, without
question or doubt, and by employing such language that these laws can be expressed in the
simplest and briefest form, they are led to think
and believe as if these laws were necessarily and
inevitably what they are. Some mathematicians
indeed have maintained that the highest laws of
nature with which we are acquainted, the laws
of motion and the law of universal gravitation, are
not only necessarily true, but are even self-
evident and certain *a priori*, like the truths of
geometry. And though the mathematical cultivator of the science of mechanics may not adopt
this as his speculative opinion, he may still be so
far influenced by the tendency from which it
springs, as to rest in the mechanical laws of the
universe as ultimate and all-sufficient principles,
without seeing in them any evidence of their
having been selected and ordained, and thus
without ascending from the world to the thought
of an Intelligent Ruler. He may thus substitute
for the Deity certain axioms and first principles,
as the cause of all. And the follower of Newton
may run into the error with which he is sometimes charged, of thrusting some mechanic cause
into the place of God, if he do not raise his views,
as his master did, to some higher cause, to some
source of all forces, laws, and principles.

When, therefore, we consider the mathema-

ticians who are employed in successfully applying the mechanical philosophy, as men well
deserving of honour from those who take an
interest in the progress of science, we do rightly ;
but it is still to be recollected, that in doing this
they are not carrying us to any higher point of
view in the knowledge of nature than we had
attained before : they are only unfolding the consequences, which were already virtually in our
possession, because they were implied in principles already discovered :—they are adding to our
knowledge of effects, but not to our knowledge
of causes :—they are not making any advance in
that progress of which Newton spoke, and in
which he made so vast a stride, in which " every
step made brings us nearer to the knowledge of
the first cause, and is on that account highly to
be valued." And as in this advance they have
no peculiar privileges or advantages, their errors
of opinion concerning it, if they err, are no more
to be wondered at, than those of common men ;
and need as little disturb or distress us, as if those
who committed them had confined themselves
to the study of arithmetic or of geometry. If we
can console and tranquillize ourselves concerning
the defective or perverted views of religious truth
entertained by any of our fellow men, we need
find no additional difficulty in doing so when
those who are mistaken are great mathematicians,
who have added to the riches and elegance of the

mechanical philosophy. And if we are seeking for extraneous grounds of trust and comfort on this subject, we may find them in the reflexion ; —that, whatever may be the opinions of those who assume the causes and laws of that philosophy and reason from them, the views of those admirable and ever-honoured men who first caught sight of these laws and causes, impressed *them* with the belief that this is " the fabric of a great and good God ;" that " it is man's duty to pour out his soul in praise of the Creator ;" and that all this beautiful system must be referred to " a first cause, which is certainly not mechanical."

2. We may thus, with the greatest propriety, deny to the mechanical philosophers and mathematicians of recent times any authority with regard to their views of the administration of the universe ; we have no reason whatever to expect from their speculations any help, when we attempt to ascend to the first cause and supreme ruler of the universe. But we might perhaps go further, and assert that they are in some respects less likely than men employed in other pursuits, to make any clear advance towards such a subject of speculation. Persons whose thoughts are thus entirely occupied in deduction are apt to forget that this is, after all, only one employment of the reason among more ; only one mode of arriving at truth, needing to have its deficiencies com-

pleted by another. Deductive reasoners, those who cultivate science, of whatever kind, by means of mathematical and logical processes alone, may acquire an exaggerated feeling of the amount and value of their labours. Such employments, from the clearness of the notions involved in them, the irresistible concatenation of truths which they unfold, the subtlety which they require, and their entire success in that which they attempt, possess a peculiar fascination for the intellect. Those who pursue such studies have generally a contempt and impatience of the pretensions of all those other portions of our knowledge, where from the nature of the case, or the small progress hitherto made in their cultivation, a more vague and loose kind of reasoning seems to be adopted. Now if this feeling be carried so far as to make the reasoner suppose that these mathematical and logical processes can lead him to all the knowledge and all the certainty which we need, it is clearly a delusive feeling. For it is confessed on all hands, that all which mathematics or which logic can do, is to develope and extract those truths, as conclusions, which were in reality involved in the principles on which our reasonings proceeded.* And this being allowed, we cannot but ask how we obtain

* " Since all reasoning may be resolved into syllogisms, and since in a syllogism the premises do virtually assert the conclu-

these principles? from what other source of knowledge we derive the original truths which we thus pursue into detail? since it is manifest that such principles cannot be derived from the proper stores of mathematics or logic. These methods can generate no new truth; and all the grounds and elements of the knowledge which, through them, we can acquire, must necessarily come from some extraneous source. It is certain, therefore, that the mathematician and the logician must derive from some process different from their own, the substance and material of all our knowledge, whether physical or metaphysical, physiological or moral. This process, by which we acquire our first principles, (without pretending here to analyse it,) is obviously the general course of human experience, and the natural exercise of the understanding; our intercourse with matter and with men, and the consequent growth in our minds of convictions and conceptions such as our reason can deal with, either by her systematic or unsystematic methods of procedure. Supplies from this vast and inexhaustible source of original truths are requisite, to give any value whatever to the results of our deductive processes, whether mathematical or

sion, it follows at once, that no new truth can be elicited by any process of reasoning."— *Whately's Logic*, p. 223.

Mathematics is the *logic of quantity*, and to this science the observation here quoted is strictly applicable.

logical ; while, on the other hand, there are many branches of our knowledge, in which we possess a large share of original and derivative convictions and truths, but where it is nevertheless at present quite impossible to erect our knowledge into a complete system ;—to state our primary and independent truths, and to show how on these all the rest depend by the rules of art. If the mathematician is repelled from speculations on morals or politics, on the beautiful or the right, because the reasonings which they involve have not mathematical precision and conclusiveness, he will remain destitute of much of the most valuable knowledge which man can acquire. And if he attempts to mend the matter by giving to treatises on morals, or politics, or criticism, a form and a phraseology borrowed from the very few tolerably complete physical sciences which exist, it will be found that he is compelled to distort and damage the most important truths, so as to deprive them of their true shape and import, in order to force them into their places in his artificial system.

If therefore, as we have said, the mathematical philosopher dwells in his own bright and pleasant land of deductive reasoning, till he turns with disgust from all the speculations, necessarily less clear and conclusive, in which his imagination, his practical faculties, his moral sense, his capacity of religious hope and belief, are to be called into action, he becomes, more than common men, liable to miss the road to truths of extreme consequence.

This is so obvious, that charges are frequently brought against the study of mathematics, as un-fitting men for those occupations which depend upon our common instinctive convictions and feelings, upon the unsystematic exercise of the understanding with regard to common relations and common occurrences. Bonaparte observed of Laplace, when he was placed in a public office of considerable importance, that he did not dis-charge it in so judicious and clear sighted a manner as his high intellectual fame might lead most persons to expect.* " He sought," that great judge of character said, " subtleties in every subject, and carried into his official em-ployments the spirit of the method of infinitely small quantities," by which the mathematician solves his more abstruse problems. And the complaint that mathematical studies make men insensible to moral evidence and to poetical beauties, is so often repeated as to show that some opposition of tendency is commonly per-

---

* A l'intérieur le ministre Quinette fut remplacé par Laplace, géométre du premier rang, mais qui ne tarda pas à se montrer administrateur plus que médiocre : des son premier travail les consuls s'aperçurent qu'ils s'étaient trompés : Laplace ne saisissait aucune question sous son vrai point de vue : il cherchait des subtilités partout, n'avait que des idées problématiques, et portait enfin l'esprit des infiniment petits dans l'administration.—*Mé-moires écrits à Ste Hélène,* i. 3.

ceived between that exercise of the intellect
which mathematics requires and those processes
which go on in our minds when moral character
or imaginative beauty is the subject of our contemplation.

Thus, while we acknowledge all the beauty
and all the value of the mathematical reasonings
by which the consequences of our general laws
are deduced, we may yet consider it possible that
a philosopher, whose mind has been mainly employed, and his intellectual habits determined,
by this process of deduction, may possess, in a
feeble and imperfect degree only, some of those
faculties by which truth is attained, and especially
those truths which regard our relation to that
mind, the origin of all law, the source of first
principles, which must be immeasurably elevated
above all derivative truths. It would, therefore,
be far from surprising, if there should be found,
among the great authors of the developements of
the mechanical philosophy, some who had refused
to refer the phenomena of the universe to a
supreme mind, purpose, and will. And though
this world be, to a believer in the Being and
government of God, a matter of sorrow and pain,
it need not excite more surprise than if the same
were true of a person of the most ordinary
endowments, when it is recollected in what a
disproportionate manner the various faculties of
such a philosopher may have been cultivated.

And our apprehensions of injury to mankind from the influence of such examples will diminish, when we consider, that those mathematicians whose minds have been less partially exercised, the great discoverers of the truths which others apply, the philosophers who have looked upwards as well as downwards, to the unknown as well as to the known, to ulterior as well as proximate principles, have never rested in this narrow and barren doctrine ; but have perpetually looked forwards, beyond mere material laws and causes, to a First Cause of the moral and material world, to which each advance in philosophy might bring them nearer, though it must ever remain indefinitely beyond their reach.

It scarcely needs, perhaps, to be noticed, that what we here represent as the possible source of error is, not the perfection of the mathematical habits of the mind, but the deficiency of the habit of apprehending truth of other kinds ;—not a clear insight into the mathematical consequences of principles, but a want of a clear view of the nature and foundation of principles ;—not the talent for generalizing geometrical or mechanical relations, but the tendency to erect such relations into ultimate truths and efficient causes. The most consummate mathematical skill may accompany and be auxiliary to the most earnest piety, as it often has been.   And an entire command of the conceptions and processes of mathe-

matics is not only consistent with, but is the
necessary condition and principal instrument of
every important step in the discovery of physical
principles. Newton was eminent above the philosophers of his time, in no one talent so much as
in the power of mathematical deduction. When
he had caught sight of the law of universal
gravitation, he traced it to its consequences with
a rapidity, a dexterity, a beauty of mathematical
reasoning which no other person could approach;
so that on this account, if there had been no
other, the establishment of the general law was
possible to him alone. He still stands at the
head of mathematicians as well as of philosophical discoverers. But it never appeared to
him, as it may have appeared to some mathematicians who have employed themselves on his
discoveries, that the general law was an ultimate
and sufficient principle : that the point to which
he had hung his chain of deduction was the
highest point in the universe. Lagrange, a modern mathematician of transcendent genius, was
in the habit of saying, in his aspirations after
future fame, that Newton was fortunate in having
had the system of the world for his problem,
since its theory could be discovered once only.
But Newton himself appears to have had no such
persuasion that the problem he had solved was
unique and final ; he laboured to reduce gravity
to some higher law, and the forces of other

physical operations to an analogy with those of gravity, and declared that all these were but steps in our advance towards a first cause. Between us and this first cause, the source of the universe and of its laws, we cannot doubt that there intervene many successive steps of possible discovery and generalization, not less wide and striking than the discovery of universal gravitation : but it is still more certain that no extent or success of physical investigation can carry us to any point which is not at an immeasurable distance from an adequate knowledge of Him.

CHAPTER VII.

On Final Causes.

WE have pointed out a great number of instances where the mode in which the arrangements of nature produce their effect, suggests, as we conceive, the belief that this effect is to be considered as the *end* and *purpose* of these arrangements. The impression which thus arises, of design and intention exercised in the formation of the world, or of the reality of *Final Causes*, operates on men's minds so generally, and increases so constantly on every additional examination of the phenomena of the universe, that we cannot but suppose such a belief to have a deep and stable foundation. And we conceive that in several of the comparatively few cases in which such a belief has been rejected, the averseness to it has arisen from the influence of some of the causes mentioned in the last chapter ; the exclusive pursuit, namely, of particular trains and modes of reasoning, till the mind becomes less capable of forming the conceptions and making the exertions which are requisite for the apprehension of truths not included among its usual subjects of thought.

1. This seems to be the case with those who maintain that purpose and design cannot be *inferred* or *deduced* from the arrangements which we see around us by any process of reasoning. We can reason from effects to causes, say such writers, only in cases where we know something of the nature of the cause. We can infer from the works of men, the existence of design and purpose, because we know, from past observation, what kind of works human design and purpose can produce. But the universe, considered as the work of God, cannot be compared with any corresponding work, or judged of by any analogy with known examples. How then can we, in this case, they ask, infer design and purpose in the artist of the universe ? On what principles, on what axioms, can we proceed, which shall

include this necessarily singular instance, and
thus give legitimacy and validity to our reasonings ?

What has already been said on the subject of
the two different processes by which we obtain
principles, and by which we reason from them,
will suggest the reply to these questions.   When
we collect design and purpose from the arrangements of the universe, we do not arrive at our
conclusion by a train of deductive reasoning, but
by the conviction which such combinations as we
perceive immediately and directly impress upon
the mind.   " Design must have had a designer."
But such a principle can be of no avail to one
whom the contemplation or the description of the
world does not impress with the perception of
design.   It is not therefore at the end, but at the
beginning of our syllogisms, not among remote
conclusions, but among original principles, that
we must place the truth, that such arrangements, manifestations, and proceedings as we be-
hold about us imply a Being endowed with consciousness, design, and will, from whom they
proceed.

This is inevitably the mode in which such a
conviction is acquired ; and that it is so, we may
the more readily believe, when we consider that it
is the case with the design and will which we ascribe to man, no less than in that which we believe
to exist in God.   At first sight we might perhaps

be tempted to say, that we infer design and purpose from the works of man in one case, because
we have known these attributes in other cases produce effects in some measure similar. But to this
we must reply, by asking how we come to know
the existence of human design and purpose *at first*,
and *at all?* What we see around us are certain
appearances, things, successions of events; how
come we ever to ascribe to other men the thought
and will of which we are conscious ourselves?
How do we come to believe that there are other
men? How are we led to elevate, in our conceptions, some of the *objects* which we perceive
into *persons?* No doubt their actions, their
words induce us to do this. We see that the
manifestations which we observe must be so
understood, and no otherwise. We feel that
such actions, such events must be connected by
consciousness and personality; that the actions
are not the actions of things, but of persons;
not necessary and without significance, like the
falling of a stone, but voluntary and with purpose like what we do ourselves. But this is not
a result of reasoning: we do not infer this from
any similar case which we have known; since
we are now speaking of the *first* conception of
a will and purpose different from our own. In
arriving at such knowledge, we are aided only
by our own consciousness of what thought, purpose, will, are: and possessing this regulative

principle, we so decypher and interpret the
complex appearances which surround us, that
we receive irresistibly the persuasion of the
existence of other men, with thought and will
and purpose like our own. And just in the
same manner, when we examine attentively the
adjustment of the parts of the human frame to
each other and to the elements, the relation of
the properties of the earth to those of its inhabitants, or of the physical to the moral nature
of man, the thought must arise and cling to our
perceptions, however little it be encouraged, that
this system, everywhere so full of wonderful
combinations, suited to the preservation, and
well-being of living creatures, is also the expression of the intention, wisdom, and goodness
of a personal creator and governor.

We conceive then that it is so far from being
an unsatisfactory or unphilosophical process by
which we collect the existence of a Deity from
the works of creation, that the process corresponds most closely with that on which rests the
most steadfast of our convictions, next to that of
our own existence, the belief of the existence of
other human beings. If any one ever went so
far in scepticism as to doubt the existence of any
other person than himself, he might, so far as
the argument from final causes is concerned,
reject the being of God as well as that of man ;
but, without dwelling on the possibility of such

fantasies, when we consider how impossible it is for men in general not to attribute personality, purpose, thought, will to each other, in virtue of certain combinations of appearances and actions, we must deem them most consistent and reasonable in attributing also personality and purpose to God, in virtue of the whole assemblage of appearances and actions which constitute the universe, full as it is of combinations from which such a suggestion springs. The vividness, the constancy of the belief of a wise and good Being, thus governing the world, may be different in different men, according to their habit of directing their thoughts to the subject; but such a belief is undoubtedly capable of becoming lively and steadfast in the highest degree. It has been entertained and cherished by enlightened and well-regulated minds in all ages; and has been, at least since the rise of Christianity, not only the belief, but a pervading and ruling principle of action of many men, and of whole communities. The idea may be rendered more faint by turning the mind away from it, and, perhaps by indulging too exclusively in abstract and general speculations. It grows stronger by an actual study of the details of the creation; and, as regards the practical consequences of such a belief, by a habit of referring our actions and hopes to such a Governor. In this way it is capable of becoming as real and fixed an impression as that of a human friend and master; and all that we can learn, by observing the course of men's feelings and actions, tends to convince us, that this belief of the being and presence and government of God, leads to the most elevated and beneficial frame of mind of which man is capable.

2. How natural and almost inevitable is this persuasion of the reality of Final Causes and consequent belief in the personality of the Deity, we may gather by observing how constantly it recurs to the thoughts, even of those who, in consequence of such peculiarities of mental discipline as have been described, have repelled and resisted the impression.

Thus, Laplace, of whom we have already spoken, as one of the greatest mathematicians of modern times, expresses his conviction that the supposed evidence of final causes will disappear as our knowledge advances, and that they only seem to exist in those cases where our ignorance leaves room for such a mistake. " Let us run over," he says, " the history of the progress of the human mind and its errors: we shall perpetually see final causes pushed away to the bounds of its knowledge. These causes, which Newton removed to the limits of the solar system, were not long ago conceived to obtain in the atmosphere, and employed in explaining meteors: they are, therefore, in the eyes of the philosopher

nothing more than the expression of the ignorance in which we are of the real causes."

We may observe that we have endeavoured to
give a very different, and, as we believe, a far
truer view of the effect which philosophy has
produced on our knowledge of final causes. We
have shown, we trust, that the notion of design
and end is transferred by the researches of
science, not from the domain of our knowledge
to that of our ignorance, but merely from the
region of facts to that of laws. We hold that, in
this form, final causes in the atmosphere are still
to be conceived to obtain, no less than in an
earlier state of meteorological knowledge; and
that Newton was right, when he believed that
he had established their reality in the solar system, not expelled them from it.

But our more peculiar business at present is
to observe that Laplace himself, in describing
the arrangements by which the stability of the
solar system is secured, uses language which
shows how irresistibly these arrangements suggest an adaptation to its preservation as an *end*.
If in his expressions we were to substitute the
Deity for the abstraction " nature" which he
employs, his reflexion would coincide with that
which the most religious philosopher would
entertain. " It seems that ' God' has ordered
everything in the heavens to ensure the duration

of the planetary system, by *views* similar to those which He appears to us so admirably to follow upon the earth, for the preservation of animals and the perpetuity of species.* This consideration alone would explain the disposition of the system, if it were not the business of the geometer to go further." It may be possible for the geometer to go further; but he must be strangely blinded by his peculiar pursuits, if, when he has discovered the mode in which these views are answered, he supposes himself to have obtained a proof that there are no views at all. It would be as if the savage, who had marvelled at the steady working of the steam engine, should cease to consider it a work of art, as soon as the self-regulating part of the mechanism had been explained to him.

The unsuccessful struggle in which those persons engage, who attempt to throw off the impression of design in the creation, appears in an amusing manner through the simplicity of the ancient Roman poet of this school. Lucretius maintains that the eye was not made for seeing, nor the ear for hearing. But the terms in which he recommends this doctrine show how hard he

---

* Il semble que la nature ait tout disposé dans le ciel, pour assurer la durée du système planétaire, par des vues semblables à celles qu'elle nous parait suivre si admirablement sur la terre, pour la conservation des individus et la perpétuité des espèces. —*Syst. du Monde*, p. 442.

knew it to be for men to entertain such an
opinion. His advice is,—

> Illud in his rebus vitium *vehementer* et istum
> Effugere errorem, vitareque *præmeditator*,
> Lumina ne facias oculorum clara creata,
> Prospicere ut possimus. iv. 823.

> 'Gainst their preposterous error guard thy mind
> Who say each organ was for use design'd;
> Think not the visual orbs, so clear, so bright,
> Were furnish'd for the purposes of sight.

Undoubtedly the poet is so far right, that a
most " vehement" caution and vigilant " pre-
meditation" are necessary to avoid the " vice
and error" of such a persuasion. The study of
the adaptations of the human frame is so con-
vincing, that it carries the mind with it, in spite
of the resistance suggested by speculative sys-
toms. Cabanis, a modern French physiological
writer of great eminence, may be selected as a
proof of this. Both by the general character of
his own speculations, and by the tone of thinking prevalent around him, the consideration of
design in the works of nature was abhorrent
from his plan. Accordingly, he joins in repeat-
ing Bacon's unfavourable mention of final causes.
Yet when he comes to speak of the laws of
reproduction of the human race, he appears to
feel himself compelled to admit the irresistible
manner in which such views force themselves
on the mind. " I regard," he says, " with the

great Bacon, the philosophy of final causes as barren ; but I have elsewhere acknowledged that it was very difficult for the most cautious man (l'homme le plus reservé) never to have recourse to them in his explanations."*

3. It may be worth our while to consider for a moment the opinion here referred to by Cabanis, of the propriety of excluding the consideration of final causes from our natural philosophy. The great authority of Bacon is usually adduced on this subject. " The handling of final causes," says he, " mixed with the rest in physical enquiries, hath intercepted the severe and diligent enquiry of all real and physical causes, and given men the occasion to stay upon these satisfactory and specious causes, to the great arrest and prejudice of farther discovery."†

A moment's attention will show how well this representation agrees with that which we have urged, and how far it is from dissuading the reference to final causes in reasonings like those on which we are employed. Final causes are to be excluded *from physical enquiry;* that is, we are not to assume that we know the objects of the Creator's design, and put this assumed purpose in the place of a physical cause. We are not to think it a sufficient account of the clouds that they are for watering the earth, (to take

* Rapports du Physique et du Moral de l'Homme. i. 299.
† De Augment. Sc. ii. 105.

Bacon's examples,) or " that the solidness of the earth is for the station and mansion of living creatures." The physical philosopher has it for his business to trace clouds to the laws of evaporation and condensation ; to determine the magnitude and mode of action of the forces of cohesion and crystallization by which the materials of the earth are made solid and firm. This he does, making no use of the notion of final causes: and it is precisely because he has thus established his theories independently of any assumption of an end, that the end, when, after all, it returns upon him and cannot be evaded, becomes an irresistible evidence of an intelligent legislator. He finds that the effects, of which the use is obvious, are produced by most simple and comprehensive laws ; and when he has obtained this view, he is struck by the beauty of the means, by the refined and skilful manner in which the useful effects are brought about ;— points different from those to which his researches were directed. We have already seen, in the very case of which we have been speaking, namely, the laws by which the clouds are formed and distribute their showers over the earth, how strongly those who have most closely and extensively examined the arrangements there employed (as Howard, Dalton, and Black) have been impressed with the harmony and beauty which these contrivances manifest.

We may find a further assertion of this view of the proper use of final causes in philosophy, by referring to the works of one of the greatest of our philosophers, and one of the most pious of our writers, Boyle, who has an Essay on this subject. " I am by all means," says he, " for encouraging the contemplation of the celestial part of the world, and the shining globes that adorn it, and especially the sun and moon, in order to raise our admiration of the stupendous power and wisdom of him who was able to frame such immense bodies ; and notwithstanding their vast bulk and scarce conceivable rapidity, keep them for so many ages constant both to the lines and degrees of their motion, without interfering with one another. And doubtless we ought to return thanks and praises to the divine goodness for having so placed the sun and moon, and determined the former, or else the earth, to move in particular lines for the good of men and other animals; and how disadvantageous it would have been to the inhabitants of the earth if the luminaries had moved after a different manner. I dare not, however, affirm that the sun, moon, and other celestial bodies were made solely for the use of man : *much less presume to prove one system of the world to be true and another false ; because the former is better fitted to the convenience of mankind, or the other less suited, or perhaps altogether useless to that end.*"

This passage exhibits, we conceive, that combination of feelings which ought to mark the character of the religious natural philosopher; an earnest piety ready to draw nutriment from the contemplation of established physical truths; joined with a philosophical caution, which is not seduced by the anticipation of such contemplations, to pervert the strict course of physical enquiry.

It is precisely through this philosophical care and scrupulousness that our views of final causes acquire their force and value as aids to religion. The object of such views is not to lead us to physical truth, but to connect such truth, obtained by its proper processes and methods, with our views of God, the master of the universe, through those laws and relations which are thus placed beyond dispute.

Bacon's comparison of final causes to the vestal virgins is one of those poignant sayings, so frequent in his writings, which it is not easy to forget. " Like them," he says, they are dedicated to God, and are barren." But to any one who reads his work it will appear in what spirit this was meant. " Not because those final causes are not true and·worthy to be inquired, being kept within their own province." (Of the Advancement of Learning, b. ii. p. 142.) If he had had occasion to develope his simile, full of latent meaning as his similes so often are, he would probably have said, that to these final causes barrenness was no reproach, seeing they ought to be, not the mothers but the daughters of our natural sciences ; and that they were barren, not by imperfection of their nature, but in order that they might be kept pure and undefiled, and so fit ministers in the temple of God.

Chapter VIII.

On the Physical Agency of the Deity.

1. WE are not to expect that physical investigation can enable us to conceive the manner in which God acts upon the members of the universe. The question, " Canst thou by searching find out God?" must silence the boastings of science as well as the repinings of adversity. Indeed, science shows us, far more clearly than the conceptions of every day reason, at what an immeasurable distance we are from any faculty of conceiving *how* the universe, material and moral, is the work of the Deity. But with regard to the material world, we can at least go so far as this ;—we can perceive that events are brought about, not by insulated interpositions of divine power exerted in each particular case, but by the establishment of general laws. This, which is the view of the universe proper to science,

whose office it is to search out these laws, is also the view which, throughout this work, we have endeavoured to keep present to the mind of the reader. We have attempted to show that it combines itself most readily and harmoniously with the doctrines of Natural Theology ; that the arguments for those doctrines are strengthened, the difficulties which affect them removed, by keeping it steadily before us. We conceive, therefore, that the religious philosopher will do well to bear this conception in his mind. God is the author and governor of the universe through the laws which he has given to its parts, the properties which he has impressed upon its constituent elements : these laws and properties are, as we have already said, the instruments with which he works : the institution of such laws, the selection of the quantities which they involve, their combination and application, are the modes in which he exerts and manifests his power, his wisdom, his goodness : through these attributes, thus exercised, the Creator of all, shapes, moves, sustains and guides the visible creation.

This has been the view of the relation of the Deity to the universe entertained by the most sagacious and comprehensive minds ever since the true object of natural philosophy has been clearly and steadily apprehended. The great writer who was the first to give philosophers a distinct and commanding view of this object,

thus expresses himself in his " Confession of
Faith :" " I believe—that notwithstanding God
hath rested and ceased from creating since the
first Sabbath, yet, nevertheless, he doth accom-
plish and fulfil his divine will in all things, great
and small, singular and general, as fully and.
exactly by providence, as he could by miracle
and new creation, though his working be not
immediate and direct, but by compass; not
violating Nature, which is his own law upon the
creature."

And one of our own time, whom we can no longer
hesitate to place among the worthiest disciples of
the school of Bacon, conveys the same thought
in the following passage : " The Divine Author
of the universe cannot be supposed to have laid
down particular laws, enumerating all individual
contingencies, which his materials have understood and obey—this would be to attribute to
him the imperfections of human legislation ;—
but rather, by creating them endued with certain
fixed qualities and powers, he has impressed
them in their origin with the *spirit*, not the
letter of his law, and made all their subsequent
combinations and relations inevitable consequences of this first impression."*

2. This, which thus appears to be the mode
of the Deity's operation in the material world,

* Herschel on the Study of Nat. Phil. Art. 27.

requires some attention on our part in order to understand it with proper clearness. One reason of this is, that it is a mode of operation altogether different from that in which we are able to make matter fulfil our designs. Man can construct exquisite machines, can call in vast powers, can form extensive combinations, in order to bring about results which he has in view. But in all this he is only taking advantage of laws of nature which already exist; he is applying to his use qualities which matter already possesses. Nor can he by any effort do more. He can establish no new law of nature which is not a result of the existing ones. He can invest matter with no. new properties which are not modifications of its present attributes. His greatest advances in skill and power are made when he calls to his aid forces which before existed unemployed, or when he discovers so much of the habits of some of the elements as to be able to bend them to his purpose. He navigates the ocean by the assistance of the winds which he cannot raise or still : and even if we suppose him able to control the course of these, his yet unsubjugated ministers, this could only be done by studying their characters, by learning more thoroughly the laws of air and heat and moisture. He cannot give the minutest portion of the atmosphere new relations, a new course of expansion, new laws of motion. But the Divine operations, on the other hand, include something much higher. They take in the establishment of the laws of the elements, as well as the combination of these laws and the determination of the distribution and quantity of the materials on which they shall produce their effect. We must conceive that the Supreme Power has ordained that air shall be rarefied, and water turned into vapour, by heat; no less than that he has combined air and water so as to sprinkle the earth with showers, and determined the quantity of heat and air and water, so that the showers shall be as beneficial as they are.

We may and must, therefore, in our conceptions of the Divine purpose and agency, go beyond the analogy of human contrivances. We must conceive the Deity, not only as constructing the most refined and vast machinery, with which, as we have already seen, the universe is filled; but we must also imagine him as establishing those properties by which such machinery is possible: as giving to the materials of his structure the qualities by which the material is fitted to its use. There is much to be found, in natural objects, of the same kind of contrivance which is common to these and to human inventions; there are mechanical devices, operations of the atmospheric elements, chemical processes;—many such have been pointed out, many more exist. But besides these cases of the combination of means, which we seem able to understand without much difficulty, we are led to consider the Divine Being as the *author of the laws* of chemical, of physical, and of mechanical action, and of such other laws as make matter what it is;—and this is a view which no analogy of human inventions, no knowledge of human powers, at all assists us to embody or understand. Science, therefore, as we have said, while it discloses to us the mode of instrumentality employed by the Deity, convinces us, more effectually than ever, of the impossibility of conceiving God's actions by assimilating them to our own.

3. The laws of material nature, such as we have described them, operate at all times, and in all places; affect every province of the universe, and involve every relation of its parts. Wherever these laws appear, we have a manifestation of the intelligence by which they were established. But a law supposes an agent, and a power; for it is the mode according to which the agent proceeds, the order according to which the power acts. Without the presence of such an agent, of such a power, conscious of the relations on which the law depends, producing the effects which the law prescribes, the law can have no efficacy, no existence. Hence we infer that the intelligence by which the law is ordained, the power by which it is put in action, must be present at all times and in all places where the

effects of the law occur ; that thus the knowledge and the agency of the Divine Being pervade every portion of the universe, producing all action and passion, all permanence and change. The laws of nature are the laws which he, in his wisdom, prescribes to his own acts ; his universal presence is the necessary condition of any course of events, his universal agency the only origin of any efficient force.

This view of the relation of the universe to God has been entertained by many of the most eminent of those who have combined the consideration of the material world with the contemplation of God himself. It may therefore be of use to illustrate it by a few quotations, and the more so, as we find this idea remarkably dwelt upon in the works of that writer whose religious views must always have a peculiar interest for the cultivators of physical science, the great Newton.

Thus, in the observations on the nature of the Deity with which he closes the " Opticks," he declares the various portions of the world, organic and inorganic, " can be the effect of nothing else than the wisdom and skill of a powerful ever-living Agent, who being in all places, is more able by his will to move the bodies within his boundless uniform *sensorium*, and thereby to form and reform the parts of the universe, than we are by our will to move the parts of our own bodies."

And in the Scholium at the end of the " Principia," he says, " God is one and the same God
always and everywhere. He is omnipresent, not
by means of his *virtue* alone, but also by his *substance*, for virtue cannot subsist without substance.
In him all things are contained, and move, but
without mutual passion : God is not acted upon
by the motions of bodies ; and they suffer no
resistance from the omnipresence of God." And
he refers to several passages confirmatory of this
view, not only in the Scriptures, but also in
writers who hand down to us the opinions of
some of the most philosophical thinkers of the
pagan world. He does not disdain to quote the
poets, and among the rest, the verses of Virgil ;

> Principio cœlum ac terras camposque liquentes
> Lucentemque globum lunæ, Titaniaque astra,
> Spiritus intus alit, totamque infusa per artus
> Mens agitat molem et magno se corpore miscet :

warning his reader however against the doctrine
which such expressions as these are sometimes
understood to express. " All these things he
rules, not as *the soul of the world*, but as the Lord
of all."

Clarke, the friend and disciple of Newton, is
one of those who has most strenuously put forwards the opinion of which we are speaking,
" All things which we commonly say are the
effects of the natural powers of matter and laws

of motion, are indeed (if we will speak strictly and properly,) the effects of God's acting upon matter continually and at every moment, either immediately by himself, or mediately by some created intelligent being. Consequently there is no such thing as the course of nature, or the power of nature," independent of the effects produced by the will of God.

Dugald Stewart has adopted and illustrated the same opinion, and quotes with admiration the well-known passage of Pope, concerning the Divine Agency, which

" Lives through all life, extends through all extent,
Spreads undivided, operates unspent."

Mr. Stewart, with no less reasonableness than charity, asserts the propriety of interpreting such passages according to the scope and spirit of the reasonings with which they are connected;* since, though by a captious reader they might be associated with erroneous views of the Deity, a more favourable construction will often see in them only the results of the necessary imperfection of our language, when we dwell upon the omnipresence and universal activity of God.

Finally, we may add that the same opinions still obtain the assent of the best philosophers and divines of our time. Sir John Herschel says, (Discourse on the Study of Natural Philosophy,

* Elem. of Phil. ii. p. 273.

p. 37.) " We would no way be understood
to deny the constant exercise of His direct
power in maintaining the system of nature ; or
the ultimate emanation, of every energy which
material agents exert, from his immediate will,
acting in conformity with his own laws." And the
Bishop of London, in a note to his " Sermon on
the duty of combining religious instruction with
intellectual culture," observes, " the student in
natural philosophy will find rest from all those
perplexities which are occasioned by the obscurity of causation, in the supposition, which
although it was discredited by the patronage
of Malebranche and the Cartesians, has been
adopted by Clarke and Dugald Stewart, and
which is by far the most simple and sublime
account of the matter ; that all the events, which
are continually taking place in the different
parts of the material universe, are the *immediate*
effects of the divine agency."

Chapter IX.

On the Impression produced by considering the Nature and Prospects of Science; or, on the Impossibility of the Progress of our Knowledge ever enabling us to comprehend the Nature of the Deity.

IF we were to stop at the view presented in the last chapter, it might be supposed that—by considering God as eternal and omnipresent, conscious of all the relations, and of all the objects of the universe, instituting laws founded on the contemplation of these relations, and carrying these laws into effect by his immediate energy, —we had attained to a conception, in some degree definite, of the Deity, such as natural philosophy leads us to conceive him. But by resting in this mode of conception, we should overlook, or at least should disconnect from our philosophical doctrines, all that most interests and affects us in the character of the Creator and Preserver of the world;—namely, that he is the lawgiver and judge of our actions; the proper object of our prayer and adoration; the source from which we may hope for moral strength here, and for the reward of our obedience and the elevation of our nature in another state of existence.

We are very far from believing that our philosophy alone can give us such assurance of these important truths as is requisite for our guidance and support ; but we think that even our physical philosophy will point out to us the necessity of proceeding far beyond that conception of God, which represents him merely as the mind in which reside all the contrivance, law, and energy of the material world. We believe that the view of the universe which modern science has already opened to us, compared with the prospect of what she has still to do in pursuing the path on which she has just entered, will show us how immeasurably inadequate such a mode of conception would be : and that if we take into our account, as we must in reason do, all that of which we have knowledge and consciousness, and of which we have as yet no systematic science, we shall be led to a conviction that the Creator and Preserver of the material world must also contain in him such properties and attributes as imply his moral character, and as fall in most consistently with all that we learn in any other way of his providence and holiness, his justice and mercy.

1. The sciences which have at present acquired any considerable degree of completeness, are those in which an extensive and varied collection of phenomena, and their proximate causes, have

been reduced to a few simple general laws. Such are Astronomy and Mechanics, and perhaps so far as its physical conditions are concerned, Optics. Other portions of human knowledge can be considered as perfect sciences, in any strict sense of the term, only when they have assumed this form ; when the various appearances which they involve are reduced to a few principles, such as the laws of motion and the mechanical properties of the luminiferous ether. If we could trace the endless varieties of the forms of crystals, and the complicated results of chemical composition, to some one comprehensive law necessarily pointing out the crystalline form of any given chemical compound, Mineralogy would become an exact science. As yet, however, we can scarcely boast of the existence of any other such sciences than those which we at first mentioned : and so far therefore as we attempt to give definiteness to our conception of the Deity, by considering him as the intelligent depositary and executor of laws of nature, we can subordinate to such a mode of conception no portion of the creation, save the mechanical movements of the universe, and the propagation and properties of light.

2. And if we attempt to argue concerning the nature of the laws and relations which govern those provinces of creation whither our science has not yet reached, by applying some analogy borrowed from cases where it has been successful,

we have no chance of attaining any except
the most erroneous and worthless guesses. The
history of human speculations, as well as the
nature of the objects of them, shows how certainly
this must happen. The great generalizations
which have been established in one department
of our knowledge, have been applied in vain to
the purpose of throwing light on the other portions which still continue in obscurity. When
the Newtonian philosophy had explained so
many mechanical facts, by the two great steps,—
of resolving the action of a whole mass into the
actions of its minutest particles, and considering
these particles as centres of force,—attempts
were naturally soon made to apply the same
mode of explanation to facts of other different
kinds. It was conceived that the whole of natural philosophy must consist in investigating
the laws of force by which particles of different
substances attracted and repelled, and thus produced motions, or vibrations *to* and *from* the
particles. Yet what were the next great discoveries in physics? The action of a galvanic
wire upon a magnet; which is not to attract or
repel it, but to turn it to the *right* and *left;* to produce motion, not to or from, but *transverse* to the
line drawn to the acting particles; and again, the
undulatory theory of light, in which it appeared
that the undulations must not be longitudinal, as
all philosophers, following the analogy of all cases

previously conceived, had, at first, supposed them to be, but *transverse* to the path of the ray. Here, though the step from the known to the unknown was comparatively small, when made conjecturally it was made in a direction very wide of the truth. How impossible then must it be to attain in this manner to any conception of a law which shall help us to understand the whole government of the universe!

3. Still, however, in the laws of the luminiferous ether, and of the other fluid, (if it be another fluid) by which galvanism and magnetism are connected, we have something approaching nearly to mechanical action, and, possibly, hereafter to be identified with it. But we cannot turn to any other part of our physical knowledge, without perceiving that the gulf which separates it from the exact sciences is yet wider and more obscure. Who shall enunciate for us, and in terms of what notions, the general law of *chemical* composition and decomposition? sometimes indeed we give the name of *attraction* to the affinity by which we suppose the particles of the various ingredients of bodies to be aggregated; but no one can point out any common feature between this and the attractions of which alone we know the exact effects. He who shall discover the true general law of the forces by which elements form compounds, will probably advance as far beyond the discoveries of Newton, as Newton

went beyond Aristotle. But who shall say in what direction this vast flight shall be, and what new views it shall open to us of the manner in which matter obeys the laws of the Creator?

4. But suppose this flight performed;—we are yet but at the outset of the progress which must carry us towards Him. We have yet to begin to learn all that we are to know concerning the ultimate laws of organized bodies. What is the principle of *life?* What is the rule of that action of which assimilation, secretion, developement, are manifestations? and which appears to be farther removed from mere chemistry than chemistry is from mechanics. And what again is the new principle, as it seems to be, which is exhibited in the *irritability* of an animal nerve? the existence of a sense? How different is this from all the preceding notions! No efforts can avoid or conceal the vast but inscrutable chasm. Those theorists, who have maintained most strenuously the possibility of tracing the phenomena of animal life to the influence of physical agents, have constantly been obliged to suppose a mode of agency altogether different from any yet known in physics. Thus Lamarck, one of the most noted of such speculators, in describing the course of his researches, says, " I was soon persuaded that the *internal sentiment* constituted a power which it was necessary to take into account." And Bichat, another writer on

the same subject, while he declares his dissent
from Stahl, and the earlier speculators, who had
referred everything in the economy of life to a
single principle, which they called the *anima*,
the *vital principle*, and so forth, himself introduces several principles, or laws, all utterly
foreign to the region of physics ; namely, *organic
sensibility*, *organic contractility*, *animal sensibility*,
animal contractility, and the like.   Supposing
such principles really to exist, how far enlarged
and changed must our views be before we can
conceive these properties, including the faculty
of perception, which they imply, to be produced
by the will and power of one supreme Being,
acting by fixed laws.   Yet without conceiving
this, we cannot conceive the agency of that
Deity who is incessantly thus acting, in countless millions of forms and modes.

How strongly then does science represent
God to us as incomprehensible ! his attributes as
unfathomable ! His power, his wisdom, his goodness, appear in each of the provinces of nature
which are thus brought before us ; and in each,
the more we study them, the more impressive,
the more admirable do they appear.   When then
we find these qualities manifested in each of so
many successive ways, and each manifestation
rising above the preceding by unknown degrees,
and through a progression of unknown extent,
what other language can we use concerning such

attributes than that they are *infinite?* What mode of expression can the most cautious philosophy suggest, other than that He, to whom we thus endeavour to approach, is infinitely wise, powerful, and good?

5. But with sense and consciousness the history of living things only begins. They have instincts, affections, passions, will. How entirely lost and bewildered do we find ourselves when we endeavour to conceive these faculties communicated by means of general laws! Yet they are so communicated from God, and of such laws he is the lawgiver. At what an immeasurable interval is he thus placed above every thing which the creation of the inanimate world alone would imply; and how far must he transcend all ideas founded on such laws as we find there!

6. But we have still to go further and far higher. The world of reason and of morality is a part of the same creation, as the world of matter and of sense. The will of man is swayed by rational motives; its workings are inevitably compared with a rule of action; he has a conscience which speaks of right and wrong. These are laws of man's nature no less than the laws of his material existence, or his animal impulses. Yet what entirely new conceptions do they involve? How incapable of being resolved into, or assimilated to, the results of mere matter, or mere sense! Moral good and evil, merit and demerit,

virtue and depravity, if ever they are the subjects of strict science, must belong to a science which views these things, not with reference to time or space, or mechanical causation, not with reference to fluid or ether, nervous irritability or corporeal feeling, but to their own proper modes of conception; with reference to the relations with which it is possible that these notions may be connected, and not to relations suggested by other subjects of a completely extraneous and heterogeneous nature. And according to such relations must the laws of the moral world be apprehended, by any intelligence which contemplates them at all.

There can be no wider interval in philosophy than the separation which must exist between the laws of mechanical force and motion, and the laws of free moral action. Yet the tendency of men to assume, in the portions of human knowledge which are out of their reach, a similarity of type to those with which they are familiar, can leap over even this interval. Laplace has asserted that " an intelligence which, at a given instant, should know all the forces by which nature is urged, and the respective situation of the beings of which nature is composed, if, moreover, it were sufficiently comprehensive to subject these data to calculation, would include in the same *formula*, the movements of the largest bodies of the universe and those of the slightest atom.

Nothing would be uncertain to such an intelligence, and the future, no less than the past, would be present to its eyes." If we speak merely of mechanical actions, this may, perhaps, be assumed to be an admissible representation of the nature of their connexion in the sight of the supreme intelligence. But to the rest of what passes in the world, such language is altogether inapplicable. A *formula* is a brief mode of denoting a rule of calculating in which numbers are to be used : and numerical measures are applicable only to things of which the relations depend on time and space. By such elements, in such a mode, how are we to estimate happiness and virtue, thought and will? To speak of a formula with regard to such things, would be to assume that their laws must needs take the shape of those laws of the material world which our intellect most fully comprehends. A more absurd and unphilosophical assumption we can hardly imagine.

We conceive, therefore, that the laws by which God governs his moral creatures, reside in his mind, invested with that kind of generality, whatever it be, of which such laws are capable ; but of the character of such general laws, we know nothing more certainly than this, that it must be altogether different from the character of those laws which regulate the material world. The inevitable necessity of such a total difference is suggested by the analogy of all the knowledge which we possess and all the conceptions which we can form. And, accordingly, no persons, except those whose minds have been biassed by some peculiar habit or course of thought, are likely to run into the confusion and perplexity which are produced by assimilating too closely the government and direction of voluntary agents to the production of trains of mechanical and physical phenomena. In whatever manner voluntary and moral agency depend upon the Supreme Being, it must be in some such way that they still continue to bear the character of will, action, and morality. And, though too exclusive an attention to material phenomena may sometimes have made physical philosophers blind to this manifest difference, it has been clearly seen and plainly asserted by those who have taken the most comprehensive views of the nature and tendency of science. " I believe," says Bacon, in his Confession of Faith, " that, at the first the soul of man was not produced by heaven or earth, but was breathed immediately from God ; so that *the ways and proceedings of God with spirits are not included in nature ; that is in the laws of heaven and earth ;* but are reserved to the law of his secret will and grace ; wherein God worketh still, and resteth not from the work of redemption, as he resteth from the work of creation ; but continueth working to the end of the world." We may be permitted to observe here, that, when Bacon has thus to speak of God's dealings with his moral creatures, he does not take his phraseology from those sciences which can offer none but false and delusive analogies ; but helps out the inevitable scantiness of our human knowledge, by words borrowed from a source more fitted to supply our imperfections. Our natural speculations cannot carry us to the ideas of ' grace' and ' redemption ;' but in the wide blank which they leave, of all that concerns our hopes of the Divine support and favour, the inestimable knowledge which revelation, as we conceive, gives us, finds ample room and appropriate place.

7. Yet even in the view of our moral constitution which natural reason gives, we may trace laws that imply a personal relation to our Creator. How can we avoid considering *that* as a true view of man's being and place, without which, his best faculties are never fully developed, his noblest energies never called out, his highest point of perfection never reached? Without the thought of a God over all, superintending our actions, approving our virtues, transcending our highest conceptions of good, man would never rise to those higher regions of moral excellence which we know him to be capable of attaining. " To deny a God," again says the great philosopher, " destroys magnanimity and the raising of human

nature; for take an example of a dog, and mark what a generosity and courage he will put on, when he finds himself maintained by a man; who, to him, is instead of a God, or *melior natura*: which courage is manifestly such, as that creature, without that confidence of a better nature than his own, could never attain. So man, when he resteth and assureth himself upon divine protection and favour, gathereth a force and faith, which human nature could not obtain. Therefore, as atheism is in all respects hateful, so in this, that it depriveth human nature of the means to exalt itself above human frailty."*

Such a law, then, of reference to a Supremely Good Being, is impressed upon our nature, as the condition and means of its highest moral advancement. And strange indeed it would be if we should suppose, that in a system where all besides indicates purpose and design, this law should proceed from no such origin; and no less inconceivable, that such a law, purposely impressed upon man to purify and elevate his nature, should delude and deceive him.

8. Nothing remains, therefore, but that the Creator, who, for purposes that even we can see to be wise and good, has impressed upon man this tendency to look to him for support, for advancement, for such happiness as is reconcile-

able with holiness;—to believe him to be the union
of all perfection, the highest point of all intellectual and moral excellence;—is in reality such a
guardian and judge, such a good, and wise, and
perfect Being, as we thus irresistibly conceive
him. It would indeed be extravagant to assert
that the imagination of the creature, itself the
work of God, can invent a higher point of goodness, of justice, of holiness, than the Creator
himself possesses : that the Eternal Mind, from
whom our notions of good and right are derived,
is not himself directed by the rules which these
notions imply.

It is difficult to dwell steadily on such thoughts.
But they will at least serve to confirm the view
which it was our object to illustrate ; namely, how
incomparably the nature of God must be elevated
above any conceptions which our natural reason
enables us to form; and we have been led to these
reflexions, it will be recollected, by following
the clue of which science gave us the beginning.
The Divine Mind must be conceived by us as
the seat of those laws of nature which we have discovered. It must be no less the seat of those laws
which we have not yet discovered, though these
may and must be of a character far different from
anything we can guess. The Supreme Intelligence
must therefore contain the laws, each according
to their true dependance, of organic life, of sense,
of animal impulse, and must contain also the

purpose and intent for which these powers were put in play. But the Governing Mind must comprehend also the laws of the responsible creatures which the world contains, and must entertain the purposes for which their responsible agency was given them. It must include these laws and purposes, connected by means of the notions, which responsibility implies, of desert and reward, of moral excellence in various degrees, and of well-being as associated with right doing. All the laws which govern the moral world are expressions of the thought and intentions of our Supreme Ruler. All the contrivances for moral no less than for physical good, for the peace of mind, and other rewards of virtue, for the elevation and purification of individual character, for the civilization and refinement of states, their advancement in intellect and virtue, for the diffusion of good, and the repression of evil; all the blessings that wait on perseverance and energy in a good cause; on unquenchable love of mankind, and unconquerable devotedness to truth; on purity and self-denial; on faith, hope, and charity ;—all these things are indications of the character, will, and future intentions of that God, of whom we have endeavoured to track the footsteps upon earth, and to show his handiwork in the heavens. "This God is our God, for ever and ever." And if, in endeavouring to trace the tendencies of the vast labyrinth of laws by which the universe is governed, we are sometimes lost and bewildered, and can scarce, or not at all, discern the line by which pain, and sorrow, and vice fall in with a scheme directed to the strictest right and greatest good, we yet find no room to faint or falter ; knowing that these are the darkest and most tangled recesses of our knowledge ; that into them science has as yet cast no ray of light ; that in them reason has as yet caught sight of no general law by which we may securely hold : while, in those regions where we can see clearly, where science has thrown her strongest illumination upon the scheme of creation ; where we have had displayed to us the general laws which give rise to all the multifarious variety of particular facts ;—we find all full of wisdom, and harmony, and beauty : and all this wise selection of means, this harmonious combination of laws, this beautiful symmetry of relations, directed, with no exception which human investigation has yet discovered, to the preservation, the diffusion, the well-being of those living things, which, though of their nature we know so little, we cannot doubt to be the worthiest objects of the Creator's care.

FINIS.